SEARCHLIGHT

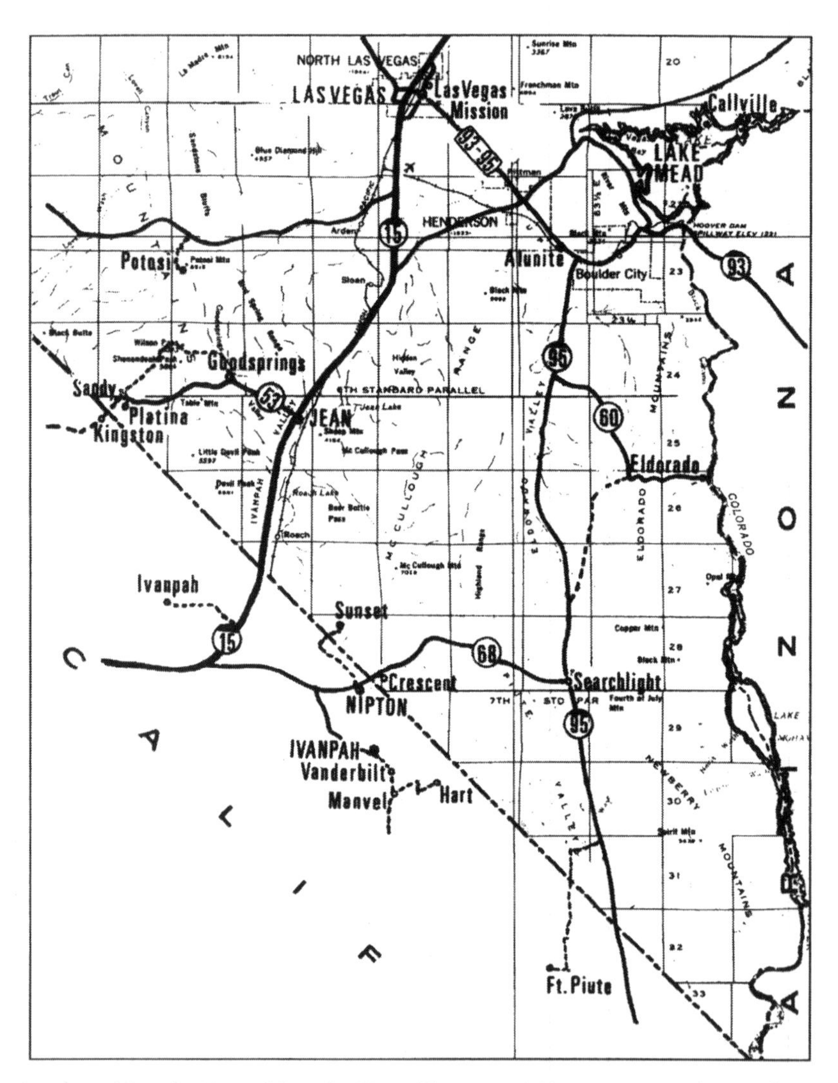

Southern Nevada. From *Nevada Ghost Towns and Mining Camps*, by Stanley Paher, 1970. Courtesy of Nevada Publications.

SEARCHLIGHT

The Camp That Didn't Fail

HARRY REID

FOREWORD BY FORMER GOVERNOR MIKE O'CALLAGHAN

INTRODUCTION BY JAMES W. HULSE

University of Nevada Press : Reno Las Vegas

University of Nevada Press, Reno, Nevada 89557 USA

Design by Carrie Nelson House
Manufactured in the United States of America
Library of Congress Cataloging-in-Publication Data
Reid, Harry, 1939–
Searchlight : the camp that didn't fail / Harry Reid ;
foreword by Mike O'Callaghan ; introduction by
James W. Hulse.
p. cm.
Includes bibliographical references and index.
ISBN 0-87417-310-8 (alk. paper)
1. Searchlight (Nev.)—History. 2. Searchlight (Nev.)—
Biography. 3. Mines and mineral resources—Nevada—
Searchlight—History—20th century.
I. Title.
F849.S43R45 1998 97-18226
979.3'13—DC21 CIP

The paper used in this book is a recycled stock
made from 50 percent post-consumer waste materials
and meets the minimum requirements of American
National Standard for Information Sciences—
Permanence of Paper for Printed Library Materials,
ANSI/NISO Z39.48-1992 (R2002). Binding materials
were chosen for strength and durability.

University of Nevada Press Paperback Edition, 2008
17 16 15 14 13 12 11 10 09 08
5 4 3 2 1

ISBN 13: 978-0-87417-753-4 (pbk. : alk. paper)

I dedicate this history of my birthplace, with fond memories,
to my Mom and Dad, Inez and Harry Reid.

CONTENTS

ILLUSTRATIONS

MAPS

FOREWORD

Searchlight, Nevada, is a strange little town, out in the middle of nowhere, in a remote corner of the world that, even today, seems somehow untouched by civilization. That a book of history should be written about Searchlight is a minor miracle. But in reading this book, one feels an immediate attraction to the town because the text is so much a labor of love, not merely an act of scholarly or academic achievement. It is clear that Harry Reid loves his hometown and that he is drawn to the sometimes strange and bewildering people who have inhabited it. Although he doesn't write about himself, he is one of those people.

For it is the people who have made this town special—from the hard-nosed miners like the author's father to the glamorous Rex Bell and Clara Bow, who came to Searchlight precisely because of its isolation. And as one learns about the place and its people, one is immediately struck by the strength of character that is essential for eking out a life in this harsh and foreboding environment. The author takes painstaking care to teach us about the people who carved out an existence in one of the loneliest outposts in America.

Unfortunately, he leaves out one of the most important stories. There was another character who was born and raised in Searchlight and who, in time, may prove to be Searchlight's most important and significant resident. And that was Harry Reid himself. Knowing him, and knowing the kind of man he is, I presume that he is just too modest to admit it.

In the future, there will be biographies written about Harry Reid, and he will be considered one of the pivotal political forces in Nevada history. A book about Searchlight would be incomplete without some mention of him and what he has done.

When I learned that this book was being published, I was immediately filled with a sense of pride, not just because the author is a good friend but also because he was one of my first students. I met Harry Reid in 1956 when I arrived in Henderson, Nevada, to take my first teaching position, at Basic High School. Harry was already in his senior year and had been elected student body president. As Harry's government teacher, and as the senior class adviser, I had daily contact with the young man who would one day become Nevada's senior senator in Washington.

That Harry was elected president of the student body might seem to some a rather startling achievement. One has to remember that, though Henderson is now Nevada's third-largest city, in 1956 it was a somewhat isolated town fifteen miles from Las Vegas that catered primarily to the working men and women who operated the Basic Magnesium plant. It was an unsophisticated, unpretentious, working-class town.

But to Harry Reid, Henderson must have seemed like New York or San Francisco. His hometown, the subject of this book, was incredibly remote; Harry was born, quite literally, in a wood shack with a tin roof, and he attended elementary school in a two-room schoolhouse. Harry loves to tell the story of how he was eager to graduate to the fifth grade so that he could go into the "big kids'" room, only to learn that there weren't enough students for two teachers and all eight grades were merged into the same classroom.

Harry's father was a hard-rock miner who spent his days deep underground with a pick and shovel, desperately carving through the rock, searching for grains of gold and other precious minerals. Harry would join him in the mines sometimes, just to keep his father company during those lonely and arduous hours. Like so many of his fellow miners, he was also a heavy drinker; life was tough and the future uncertain, and the potential for life-threatening accidents lay around every corner and with every blasting cap. He eventually committed suicide after a severe bout with depression.

Inez Reid, Harry's redheaded mother, was a strong, clear-eyed woman with little formal education. Obviously, it was a hardship raising four sons in such harsh surroundings, but she had a wistful, optimistic spirit that was handed down to her children. I knew her only later in life, after she had been ravaged by the years of hard work and toil, but her spirit was still unwavering.

Searchlight was so far removed from urban life that when Harry started school in Henderson in 1953, he was forced to hitchhike the forty miles into town each Monday and stay with relatives during the week, hitchhiking back home to be with his own family on the weekends. He must have seemed forlornly out of place, a rural boy in an industrial town.

Harry was shy and awkward, and probably befuddled by the strange vagaries of "urban" life. By 1956, however, he had grown accustomed to his new environment and had become popular in the school. Harry was a determined young man; though he was not a great athlete, he loved sports and played on the school's football and baseball teams. He became good through tenacity. Harry was not the biggest or the fastest player, but he displayed grit, eagerness, and a fierce competitive spirit that more than made up for his physical limitations.

Harry even took up boxing at the Henderson Boys Club, where I was the boxing coach. As a training technique, I made the boxers run up the steep grade from Henderson to Railroad Pass, following them in my car. They knew that if they stumbled or fell and did not promptly get up and continue, there was a good chance they would be run over. Harry never complained. In the ring, he shook off adversity, especially when matched against a superior opponent, and he went on to become a good middleweight. His unflagging drive to succeed made him a champion, and this tenacity has carried through to his private and public life as well. His stamina and energy are also reflected in his participation in several marathon races, including the Boston Marathon, and he continues to run on a regular basis despite the grueling schedule of the Senate.

Harry was also one of my finest students. There was always something different about Harry, however; and I think, now, that it must have been the spirit of the mines in Searchlight, something raw and untamed and confident. He had no fear.

I might add that he was also dating one of the prettiest girls in the school, Landra Gould. As one of the most popular girls in school, Landra could have had her pick of the boys, and some were amused that she chose Harry. Her father, a prominent physician in Henderson, was especially opposed to their relationship. Obviously, Landra saw in the young Harry what the rest of us would learn in due time—even her father would eventually come around, warmly accepting Harry as his son-in-law.

When Harry graduated from high school, a group of local business people contributed to a fund to help send him to college. Harry chose Southern Utah State College in Cedar City, where he had earned an athletic scholarship. This was a kind of precursor to student loans, which would be repaid when Harry returned to Henderson. We remained in close contact while he finished his undergraduate degree at Utah State University and after, when he and Landra moved to Washington, D.C., where Harry attended the George Washington University School of Law. As a law student, Harry struggled with the difficulty of raising a young family while also working nights as a Capitol policeman.

I can still remember meeting Harry at the Reno airport on a chilly autumn evening when, during the middle of his final year of law school, he returned to Nevada to take the Nevada bar exam. He was broke, and I talked him into taking a few dollars from me to help with food and lodging on his quest to become a lawyer. Harry is one of the few people who have successfully passed the Nevada bar before completing law school. Harry repaid all of those school loans, and there is no one who regrets having invested in this young man's future. Little did I know that after his bar examination our paths would continually cross.

When Harry and his family returned to Henderson, his adopted hometown, he was hired as Henderson's city attorney and he also began a successful practice as a trial attorney. In 1966 he ran for his first public position, as a member of the Southern Nevada Hospital board of directors; two years later, he was elected to the Nevada State Assembly. In 1970, while running for governor of Nevada, I had the good fortune of having Harry Reid run as lieutenant governor. Many people thought that Harry ran on my coattails; what they didn't realize was that Harry received more votes than I did. Regardless, in November 1970, at age thirty, Harry was elected as the youngest lieutenant governor in the state's history.

I could not have asked for a more trustworthy or capable assistant. During the state's biennial legislative sessions, Harry was my chief legislative strategist and vote counter, and I credit him for the many successes we had in the legislature. Harry had total and complete access to any meeting in my office or at the governor's mansion, and he also was the person who arranged the historic meeting between Howard Hughes and me in London. As governor, I was in some ways still Harry's mentor, but he took those lessons to new heights.

In 1974 Harry ran for the U.S. Senate seat vacated by the legendary Alan Bible. Throughout the race, Harry led former governor Paul Laxalt in the polls; however, on election day, he fell short by a mere 600 votes. He was obviously devastated by the defeat, but he took the loss in stride and returned to southern Nevada to build his law practice and devote himself to his family.

I was extremely proud of Harry when he accepted his most courageous and, perhaps, most controversial civil case, representing a former classmate in a discrimination suit against the Clark County Sheriff's Department. The plaintiff, Larry Bolden, was a star athlete and a talented student at Basic High School, and one of the first black policemen hired in Nevada. He had risen to the rank of sergeant but was denied promotion to lieutenant, even though his scores on the exams clearly established his qualifications. Harry took that case and brilliantly exposed the racial discrimination within the police department during a time when such arguments were not welcome or encouraged. The case was won, and Larry was promoted; he eventually earned the rank of deputy chief and became second in command of the state's largest police force.

I tell this story because it illustrates how Harry never backs off from a challenge when he knows that, in principle, it is the right thing to do. It was for this reason that I appointed him chairman of the state's most important board, the Nevada Gaming Commission. To this day, I believe that was my most significant appointment. Harry led the commission through its most challenging time, confronting the influence of organized crime in Nevada gaming and ushering in a new era of corporate involvement in Nevada's resort economy.

In 1982 Harry again answered the call to public service by running for congressman in Nevada's First Congressional District. He was elected and had served two terms when, in 1986, he ran for the Senate seat vacated by Senator Paul Laxalt, the same seat he had come so close to winning twelve years earlier. This time, however, Harry was the underdog, and I think that may have made the difference. Just as when he was a boxer, Harry was more tenacious, more spirited, and tougher when the odds were against him and when he was faced with a stronger opponent. Again he rose to the challenge, mounting a perfect campaign and proving all of the pundits wrong.

Since then, Harry has taken the role of the underdog to unprecedented levels. As a first-term senator, he battled against the odds to create Nevada's first national park, to protect wilderness lands, to negotiate successfully a Truckee/Carson River water settlement that ended more than eighty years of water wars, and as a member of the powerful Appropriations Committee, he has been able to support the infrastructure needs of the country's fastest-growing state.

As a national legislator, he has authored the Taxpayers Bill of Rights, sponsored legislation outlawing the infamous source tax, passed the Child Victims Bill of Rights, outlawed the ugly practice of female genital mutilation, sponsored lead abatement legislation, and much more.

His style as a behind-the-scenes negotiator has become so legendary that he was selected by his peers to serve in a leadership role in the U.S. Senate. He is now a recognized expert on environmental protection, mining reform, western land issues, clean air and safe water law, domestic policy, and a host of other concerns. And in 1993 *Parade* magazine listed Harry as one of six senators known for their character and integrity.

Little did his teachers in Searchlight know that, in performing their daily tasks, they were preparing a young boy for a major role in national politics. I don't know where Harry's political future will take him, but I do know that the lessons he learned in Searchlight—lessons about handling adversity, confronting a harsh and extreme environment, and facing challenges with courage and principle—will provide the foundation for his life's work. That's why this historical tribute to Searchlight is so important; it gives insight into the man who wrote it.

I barely knew Harry's father, and I knew his mother only briefly, but I know they would have been proud of the accomplishments of the son who would be senator. So are all of us who have watched Harry Reid's development.

As this book attests, Harry Reid is proud of his roots in Searchlight. Searchlight can also be proud of him.

FORMER GOVERNOR MIKE O'CALLAGHAN

PREFACE

I was born in Searchlight just before the outbreak of World War II, but the Reids' family history in Nevada went back many years before my birth. My paternal grandparents were listed in the census and voting records of 1900 and 1910. My grandfather, John Reid, is mentioned several times in early Searchlight newspapers.

The Reids left Searchlight about 1910, after the ore failed, but returned in the late 1920s, when there was hope that the mines would again become productive. My father was a hard-rock miner, and I often went underground with him. I learned to love not only the desert aboveground but also the world below.

My three brothers and I all attended Searchlight Grammar School. My formal education began there in the first grade and ended upon my graduation from the eighth grade at Searchlight Elementary School. There was no more school to be had in Searchlight; the town was still too small to support a high school. An aspiring student went to high school wherever he or she could, so I left Searchlight in the autumn of 1952 to attend Basic High School in Henderson, some forty miles from home. This was the end of my full-time residence in the place of my birth.

During the first of my four years at Basic High, I came home on weekends. After that, I came home less frequently. During college, I came to Searchlight only on rare occasions, and while in law school in the East, I was unable to visit Searchlight at all.

Even though I began to spend less and less time in my hometown, my thoughts often returned to the days of my youth. Most of all, I realized how much I loved the desert. When I completed my education, I lived in Las Vegas. My parents still lived in Searchlight, and their four boys visited them and other family members there as often as possible.

I always had questions about the formation and development of Searchlight, but there was no complete and accurate history of this fascinating town to consult. In recent years I became increasingly curious, but like many historians, I waited far too long to begin a serious study of Searchlight's past. I have worked on this history for eighteen months. During that time, four early residents whom I interviewed have died. But for this book, their insightful perspectives on the town's history would have been lost, like those of many who have passed before them. This book would have been better if I had started earlier, before so many of the details, such as dates and locations of important events, were lost.

The evidence that reveals the history of Searchlight did not come easily. Newspapers were the key in accumulating information, especially about the first two decades of the town's existence. I will always be indebted to an elderly woman named Arda Haenszel, who published recollections and reminiscences of her school days in the fading mining town. Magazine pieces and newspaper articles from papers outside of Searchlight, though not always accurate, were probative and many times provided important information that was available nowhere else. Some benefit was derived from court records involving facts and figures of the town. And when available, historical papers from individuals and companies proved to be valuable sources of information. Interviews with those who had firsthand knowledge of Searchlight provided magnified information about the town and its people. These conversations produced hundreds of photographs and several valuable letters.

As a student of history, I have enjoyed the research. Time spent on this project brought back many memories for me. In a broader sense, the preparation of this book also indelibly impressed upon me the importance of historical preservation.

Many have helped make this book possible. Because this has been a part-time venture for me, I have had to conduct much of the work on weekends and on airplanes; I even spent most of Christmas 1995 working on the manuscript. I am, therefore, most indebted to my family, especially my lifetime companion, Landra, for their patience and understanding.

Dennis Casebier, a historian of the Mojave Desert, has rendered unselfish assistance to me. He shared his extensive library, conducted more than a dozen interviews, and provided encouragement to me to finish the project. The book is better because of Dennis.

My oldest brother, Don, was generous with his time, as he has always been. He conducted several recorded interviews, generating a better insight and understanding of Searchlight. My youngest brother, Larry, was a valuable asset because of his incredible knowledge of the geography of Searchlight and the surrounding territory. His memory of the previously unwritten history of the place of our birth was better than mine. My son Josh worked hard preparing the bibliography, and he also assisted me in selecting the appropriate maps.

My friend of more than twenty years, Curt Gentry, who has written many books of his own, including several best-sellers, has been a tremendous asset to me because of his interest in history and writing.

Professor Andrew Tuttle, of the University of Nevada at Las Vegas Political Science Department, was also an ardent supporter of this enterprise. His research on Queho was penetrating and thorough.

One of my high school friends and baseball teammates, later to become a law associate, J. Bruce Alverson, enrolled in some graduate history courses at UNLV at the same time I was beginning my book, and unknown to me, he chose to write a research paper on Searchlight. Even though the paper was less than thirty pages long, Bruce's research, especially on early Searchlight newspapers, was very detailed. He turned these research materials over to me, and they became an important resource for the early chapters of this book.

Some Searchlight old-timers were also most helpful. Donna Jo Andrus, now a resident of Nelson, was very generous with her time and resources and was a reservoir of materials. Her friend and mine for more than fifty years, Joyce Dickens Walker, was enthusiastic in her help, mailing a package of photos and accompanying narratives at least once a month. This material was essential for a better understanding of Searchlight.

Searchlight has its own resident historian in Jane Overy. She started the Searchlight museum and has worked hard to collect memorabilia pertaining to the town. She was kind and thoughtful in reviewing my first draft, and her input was very valuable.

Sally Macready Wallace, the granddaughter of one of the town's true pioneers, Benjamin Macready, and the daughter of one of the town's heroes, John Macready, was exceptionally forthcoming with information that she has accumulated about her famous forebears. She has formed the Macready Foundation, and she shared many of her intellectual treasures with me.

Many others were helpful also. Pierre Valentini, a family friend, became an expert at dating early Searchlight newspapers. Early research identified important news articles, which were then marked and copied but not dated. Pierre undertook the tedious task of matching articles with dates.

No one has been more helpful than William Marion. He gave me the benefit of his wisdom in outlining the available resources for the research project, and his final review of my last draft showed his brilliance in the use of the English language. His input greatly helped my output.

Jeri Pickett Werner devoted many hours to typing and retyping various drafts of the book. She made my life much easier and also gave me an opportunity to get away from the book, since for several weeks it became her problem.

Most of all, I extend my gratitude to those who gave me an enthusiasm for history. Former Nevada governor Mike O'Callaghan, my high school history teacher, inspired me in many areas of life, but none more than a love of history. Professor Leonard Arrington, eminent scholar, historian, and my teacher at Utah State University, made it clear why even a small town like Searchlight has historical significance that demands preservation.

As Justice Benjamin Cardozo wrote, "History, in illuminating the past, illuminates the present, and in illuminating the present, illuminates the future." It is my hope that this first history of Searchlight truly will illuminate the past and, in so doing, allow us all to better focus on the future.

INTRODUCTION

In the office of Senator Harry Reid of Nevada on Capitol Hill in Washington, a conspicuous piece of furniture is a spacious conference table covered with a glass top. Beneath the glass is a large aerial photograph of Searchlight, Nevada, the senator's hometown and the place that most genuinely claims his affection. The photograph covers the entire table and is one of the most prominent objects in the room.

There is little to recommend the tiny town to the attention of the world beyond the senator's office. Situated fifty miles south of Las Vegas in one of the least inviting parts of the Mojave Desert, Searchlight retains few reminders of the glory days of ninety years ago when it was one of the names to be reckoned with among the promising gold mining camps of the Southwest. Even during its most prosperous period, soon after the turn of the century, it had little beyond its mines to invite commercial interest. The high-altitude photograph in the senator's office may have been the most visible representation available until the publication of this book.

This unpretentious community, which has been almost totally neglected by the chroniclers of Nevada and the Southwest, was one of several hard-rock camps that appeared on the regional landscape between 1895 and 1910. Its ores were discovered shortly before the turn of the century, during a time of depression in the precious metals industry. The first wave of speculators established a traditional mining district in 1898, and the first notable production of gold was recorded in 1899. Searchlight remained on the roster of hopeful camps for slightly more than a decade, reaching its peak in 1906 and 1907, then suffering the familiar loss of confidence and population that was the fate of so many outposts of the mining frontier. It was never totally abandoned, however, and experienced a brief revival in the 1930s, after New Deal policies boosted the price of gold. This resurgence reawakened the hopes of a handful of die-hard prospectors for yet another generation.

Before beginning the main text here, the reader might wish to pause momentarily to reflect on the southern Nevada setting at the turn of the century, just after Searchlight had attracted its first wave of prospectors. In 1900, only 42,335 people lived in the entire state, and virtually all of them made their homes in the northwest, around Reno and Carson City or along the route of the Central Pacific Railroad on the Humboldt River. Southern Nevada was almost totally undeveloped.

Sprawling Lincoln County, in which Searchlight was located, was then the largest in Nevada, covering 18,700 square miles. It had, however, only 3,284 residents, according to the official census of 1900. More than 900 of these lived in Delamar, in the Meadow Valley Mountains 150 miles north of Searchlight; that was its nearest mining-camp neighbor within Nevada. Gold had been discovered there in 1894, and for several years more than $1 million annually had been extracted from its underground workings. The relative success of Delamar during the depression years of the middle 1890s was one inducement for prospectors to explore more diligently the outcroppings in the southern terrain, which had been probed intermittently but unsuccessfully in the past. The earliest reports of the Searchlight discovery appeared in the *DeLamar Lode* and the *Pioche Record,* newspapers hungry for reports of good prospecting possibilities anywhere in the county. Pioche, the county seat, was 225 miles north of Searchlight; a trip to do business there required as much as ten days of hard travel over harsh terrain. No town in Nevada was more remote from its county seat and the state capital. (While there were disadvantages in having basic governmental services so far away, on the other hand the potential nuisances of mine inspector, county assessor, tax collector, and sheriff were also quite remote.)

When Searchlight made its appearance on the mining horizon, the Territory of Arizona was still awaiting admission to the Union as a state, and the Las Vegas ranches had a population of 30 people, as recorded by the census of 1900. In that year the enumerators counted 211 residents in Searchlight.

Yet, as Reid's book indicates, Searchlight was not quite as isolated from the avenues of commerce as the foregoing recital of facts would suggest. The Colorado River, which was navigable to the south, is only fourteen miles to the east. Wagon roads had been established through the region much earlier. Needles, California, is forty-five miles to the south, and through that town passed the Atchison, Topeka, and Santa Fe Railroad, built along the 35th parallel in the early 1880s. Although forty-five miles is not an insignificant distance in this arid terrain, it was less forbidding than the vast expanses that many earlier Nevada camps further north had endured at their inception. Reid reminds us that during the period of Searchlight's greatest commercial activity in 1906, a connection was developed with the San Pedro, Los Angeles, and Salt Lake Railroad, which by that time passed only twenty-three miles to the west.

Those who know the history of the far western mining frontier will find much that is familiar in this narrative. The rapid rise to prominence, the industrial trouble between mine owners and laborers that occurred early in the camp's history, the speedy introduction of many modern technical and commercial amenities, and the absence of a church until late in its history—all these characteristics will be recognized by the devotees of mining history. "The Camp Without a Failure," which was the logo of Searchlight's local newspaper, faded less rapidly than most of its ephemeral counterparts to the north. Its afterglow enabled it to survive into the era of the Great Depression (when Harry Reid was born there) and into its centenary year—1998—by which time the real estate boom had transformed its environs once more.

Senator Reid obviously enjoyed appending the ironic subtitle. In a sense, Searchlight was "the camp that didn't fail" because it was never really a great success as a mining town in the context of Nevada history, except in the anticipation of those who lived there and bonded to it.

Appearing on the mining maps of the West half a century after the California gold rush and four decades after the opening of the great Comstock Lode in northwestern Nevada, it generated a flurry of interest in mining venture circles, but by that time investors were chary of reports of new bonanzas.

During the decade of its highest productivity, 1900 to 1910, Searchlight had to share the spotlight with such boomtowns as Tonopah and Goldfield, also more than 200 miles away but much more prominent competitors than Pioche or Delamar. In the year of Searchlight's greatest recorded gold ore production—1906—official records show that its output was about $590,000, whereas Goldfield reported more than $6.6 million and Tonopah more than $3.4 million. It is likely that the actual production was substantially higher for each district, because it was not unusual for mining companies to underreport their production to avoid the state tax on the net proceeds of mines.

Upon reaching the part of this narrative that describes the "big strike," the reader may recall that the decade of the 1890s was a tense period in the history of industrial relations in the mining camps of the West. The action of the Searchlight mine owners in 1903 in defying a newly enacted state law relating to working hours reflected tension that had existed for most of the previous decade. The events that Reid describes here indicate that those who owned the mines in Searchlight decided they had to take preemptive action early to discourage the kind of labor unrest that had erupted in mining towns in the Rocky Mountains.

The election of 1906 brought 216 voters to the polls in Searchlight in the race for the congressman from Nevada. By comparison, 155 votes were cast in Las Vegas and only 110 in Pioche. For a few brief seasons thereafter, Searchlight was the largest town in Lincoln County.

Reid's reconstruction of the evolution of this typical gold camp at the turn of the century enables us to contemplate an era and a place in which government, transportation, communication, and commerce were only slightly more advanced than they had been in the early years of the nation's history. Yet the pioneers of 1900—and especially those who followed the mining frontier—were much more mobile and often more willing than their grandfathers had been to endure a hostile environment in search of the magnetic "yellow metal."

Searchlight was born too late to reap some of the benefits that existed for its predecessors that had appeared in Nevada a generation earlier. It never had the advantage of becoming a county seat, as a few of the older and more fortunate of its northern peers did. By 1909, when the legislature formed Clark County from the southern part of Lincoln and established a county seat in Las Vegas, Searchlight was already declining.

Much of the history of these remote places on Nevada's mining frontier vanished with their fleeting prosperity either because they did not have newspapers to record their hopes, successes, and failures or because, when they did, copies did not survive. Searchlight and Reid have both been fortunate in that regard. Many short-lived bonanza towns have been celebrated by writers of the Sunday newspaper features or romanticizing storytellers, but only infrequently have they undertaken the dedicated research in the periodical press and other sources that is necessary for serious history. Searchlight had a weekly newspaper for slightly more than a decade, from 1902 until 1913, and almost the entire run is preserved in Nevada historical libraries. The senator used these files diligently to reconstruct his account of the community's life. He also scoured the files of the later ephemeral local papers that appeared from time to time.

The populations of these camps were notoriously transient; it was common for a miner to linger at most only a few months in a place like Searchlight when the lure of other camps was greater. The tenacity of a family like the Reids, who arrived in the town early in its history, departed for more promising sites, and returned, is unusual.

One of the singular virtues of this book is its panoramic view of a distinctive boom-and-bust camp *after* its "palmy days," as the old-time miners and newspapers used to call the better years of their memory. As the son of a miner who prospected and kept the faith long after most others had moved on, Reid developed a keen affection and an insatiable curiosity that he did not lose when he moved upward and onward in public service.

In this book we have a chronicle of those who lived here but who left early in life to become famous, those noteworthy folks who passed through, and those who simply made use of its name. Of the celebrities who made a home in the region during the long Indian summer of Searchlight's history, the early movie stars Rex Bell and Clara Bow were among the most interesting. They were past their prime in Hollywood, but they retained a penumbra of fame in the 1940s and found their place of sanctuary on a nearby ranch. Bell became a prominent Nevada political figure and also served as a lieutenant governor of Nevada. We also get profiles of others who attained prosperity and prominence after they left these parts.

Reid offers us here a cornucopia of data and anecdotal material about Searchlight and its environs. Such subjects as the problem of a water supply in the parched Mojave, the earliest efforts to operate an electrical generating plant, the technical details of the mineral-extracting industry—all come under Reid's scrutiny. The arrangement of his material is occasionally eclectic, somewhat in the manner of an old-timer reflecting on those earlier days of success, but the fact that he extracted such details and put them at the disposal of historically minded readers is evidence of his prospector-like tenacity. He has gathered nuggets of Searchlight history the way an unrehabilitated prospector gathers samples of ore to show his friends.

During his career in politics, the senator has gained the respect of active environmentalists in Nevada as well as many leaders in the minerals industry, who have often been at odds with each other. That he came to love his desert surroundings and at the same to appreciate the ethos of the miners obviously helped him reach this accommodation.

By coincidence, as the manuscript of this book was undergoing its final revision in preparation for publication, an article appeared in the *Nevada Historical Society Quarterly* (Fall 1996) on the history of Searchlight between 1903 and 1909. Its author was Bruce Alverson, a graduate student at the University of Nevada, Las Vegas. Alverson's article does not in any way diminish the value of Reid's much more extensive treatment of this corner of history. In fact, it might well serve as an appropriate introduction to the present work. There are not likely to be any charges of "claim jumping," even though each author has extracted ore from the same veins of Searchlight's history.

A closing personal note may be in order. I feel a kinship with Harry Reid because I was born and came of age in Pioche, that remote county seat and another historically neglected mining town. I, too, followed my father into the mines and the mountains at an early age, looking for the signs that would guide us to the anticipated but elusive bonanza. Most of us sons of the compulsive prospectors got out of that business early in life and went to the cities, but we have inherited some of the romance of our memories. One can find in these pages echoes of the unrealistic conviction that this was the greatest camp of them all and that somehow in our remembrance it was a better place than the historians who ignore it have been willing to acknowledge. Most history is written about the more enduring success stories, and mining history too concentrates on the places that "made the big time." It may be argued by the commentators that Searchlight was not as great a camp in its prime as Reid's narrative would encourage us to believe. On the other hand, its story embraces a kind of experience in the development of the Far West that was as much a part of the historical tapestry as those accounts of the more frequently chronicled towns.

Who will prove that Searchlight was a failure? Towns, like individuals, have pride that their native sons may carry along with them in later years despite the empirical evidence. With this book and his career of public service, Senator Reid has validated the assertion that is implicit in his title, even if the economic evidence that he has assembled does not confirm it. That aerial photograph in his U.S. Senate office is a wry verification that the newspaper slogan was more apt than we realized.

JAMES W. HULSE
UNIVERSITY OF NEVADA, RENO

1

THE BEGINNING

Searchlight is like many Nevada towns and cities: it would never have come to be had gold not been discovered. Situated on rocky, windy, and arid terrain without artesian wells or surface water of any kind, the place we call Searchlight was not a gathering spot for Indian or animal.

Only fourteen miles to the east is the Colorado River. Ten miles to the west is a modest mountain range, with fragrant cedars, stately pines, and a few sheltered meadows, home to an ancient Indian camp referred to as Crescent. To the northeast lies the canyon called Eldorado. In the eighteenth century the Spaniards explored and then mined this area.[1] The same location was exploited by Brigham Young, who directed some of his Mormon followers to present-day Nevada in search of minerals for his Utah civilization.[2] To the southwest, about fifteen miles distant, is the site of a U.S. military frontier outpost, Fort Piute or Piute Springs.

The mighty Colorado River was used for various routes along the navigable portion of its course. The main impediment to through passage from the north was the Grand Canyon, but the river was usable for about a hundred miles above Searchlight to as far south as the border of present-day Mexico.

During the Civil War the U.S. military tried to find better routes for moving men and supplies. Captain George Price, who had been commissioned by his superiors to find an easier route from the area of Salt Lake City to the southern part of the Utah Territory, led one such effort.[3] He left Camp Douglas, near Salt Lake, on May 9, 1864, and worked his way south to Fort Mojave, near what is now Laughlin, Nevada. The trip was uneventful until he reached present-day Cedar City, Utah. The route over the desert from there to Las Vegas was extremely harsh and inhospitable. From Las Vegas to Eldorado was easier, but the journey from Eldorado to Fort Mojave was particularly brutal. The route then proceeded to Lewis Holes, an area west of Piute Springs named after Nat Lewis, an early Eldorado Canyon miner.[4] After arriving at Fort Mojave, Captain Price declared that the route was unsafe and unsuitable for military use.

As an interesting note, during Price's journey his company came upon a stray cow at a watering spot near Lewis Holes and a place called Government Wells. Price's men killed and ate the cow, and the watering hole was formally named Stray Cow Wells in recognition of the event.[5]

The accepted route that Captain Price and others traveled was called the Eldorado Canyon Road, which went from Eldorado Canyon to the Lanfair Valley and wound its way through the Castle Mountains, ending at Lewis Holes. Many prospectors traveled over the road, but written accounts have focused on the conditions of travel rather than describing the trail itself.[6]

This pioneer route came very close to present-day Searchlight. As Dennis Casebier points out in his *Mojave Road Guide,* "Eldorado Canyon is usually a dry side canyon coming in to the Colorado River from the west about 25 miles below Hoover Dam. The route to the mines in the Canyon from Los Angeles took the Mojave Road to this point. From here the road angled off to the northeast via Lewis Holes toward the present Searchlight, then turned northward to Eldorado Canyon. Connections were developed from the Eldorado Canyon to Las Vegas and the main Salt Lake Trail. This point was a major road junction of the day. Here travelers had to decide whether to go northeast toward Utah or continue directly east on the Mojave Road toward Arizona and New Mexico. This intersection fulfilled the same purpose as the present junction of I-15 and I-40 in Barstow, California."

Eldorado Canyon was the object of Anglo exploration long before Brigham Young's forays and the U.S. Army's expeditions, however. Clearly, the first white man to pass through or near Searchlight was Father Francisco Garcés in 1776. He left no physical sign of his passing, but his journals are sufficiently detailed to indicate that he came near the town.[7]

Several of the mines in Eldorado Canyon have a long unwritten history that some believe goes back two centuries. Even though there is no written account of any Spanish or Mexican mining enterprise in the canyon, it is clear that such activity did take place. John Townley reports that mining likely went on there between 1750 and 1850. The mining operations never spilled over into Searchlight, but the explorations came very close.[8]

From its earliest days, Searchlight had significant interaction with Eldorado Canyon. By the time Searchlight was founded, Eldorado had long been in operation. The contact was closest before the railroad came to Searchlight, when the mines and the people depended more on the river. The landing at the mouth of Eldorado Canyon was more important to the mines, however, than the river at Cottonwood was to Searchlight.

Reports like the following from a conversation with John Riggs contrast the operations in Eldorado and Searchlight: "John Powers, who is still living and who at one time owned the Wall Street Mine, told me one evening about 1882 that an outfit of Mexicans of the better class rode up to his camp at the Wall Street, and asked him if he owned the mine. He replied that he did. They then said that they had a very old map of this country and that the Wall Street was marked on the map. The map was evidently correct as they had come straight to the mine. They stated that the map had been made very long ago, probably by early Spaniards."[9] The Wall Street was one of the big producers of gold in Eldorado Canyon for many years. Conversely, no mine in Searchlight, with perhaps the exception of the Quartet, was worked successfully for more than ten years.

Though we do not know when the activity in Eldorado Canyon actually began, we do know that the mining district had a hectic and eventful history in the latter part of the nineteenth century. One account puts as many as 1,500 people there during the Civil War.[10]

The first documented records of contemporary mining in the Searchlight area were provided by a mining company called Piute, which was formed in 1870. This company owned 130 mines in California and in southeastern Nevada. The most prominent of the Nevada mines was the Crescent, located about ten miles west of Searchlight. The company's promotional documents described a road that passed near present-day Searchlight and went to Cottonwood Island, below Searchlight on the Colorado River. The road was said to be favorable, with a broad, smooth path, much of it along a dry ravine.[11]

In the early 1870s, a promoter named Johnny Moss attempted to develop a city just off Cottonwood Island. The town, which would be called Piute, was to be the freight head for the mines headquartered at Ivanpah, some forty miles to the west. The project never went beyond an artist's rendering, however. The proposed mines were later developed, but San Bernardino rather than Ivanpah emerged as the shipping terminus.[12]

Indians traveled from the mountains above Searchlight to the river, creating relatively extensive foot traffic near the town's present location, and miners passed through the area in their never-ending quest for the gold and silver of their dreams.

When Searchlight was established at the end of the nineteenth century, the mining camp with the unusual name had a very primitive infrastructure, but it swiftly became modern. Within a few years Searchlight was as fashionable as any western town of its day. Its amenities were noticeably contemporary. A modern water system was quickly created, incorporating pumping facilities, a new storage tank, piping, fire hydrants, and meters. The town even had a telephone system, which for the time was very advanced, and a telegraph system. An outdated railroad was soon replaced by a more modern line that included passenger travel. Surprisingly, early Searchlight had a modern system of electricity and its own power plant.

The places of business in town were many and varied, including a barbershop, several saloons and hotels, a lumberyard, clothing stores, sundry shops, cafes, union halls, boardinghouses, schools, garages, and stables. The town even boasted a hospital with doctors and, of course, a newspaper or two.

When the mines' production waned after 1908, the businesses slowly began to cut back and in many instances simply failed. The decline, though sporadic, was technologically regressive. By the late 1940s and 1950s there was very little left of the modern Searchlight. Fires and a lack of prosperity had ravaged the once thriving community, and now there were no barbershops, no hotel, no lumberyard, no clothing store, no sundry shops, no union hall, and not even the trace of a union. Of course, the need for a hospital had long since ceased. There was no doctor, not even on a part-time basis.

In the town's early days, especially with the coming of the railroad, the grocery stores carried a full line of food and merchandise. Fresh produce came from the farms around the area, including the river and Lanfair Valley, and beef came by rail, stage, and truck, as well as from the nearby ranches. Near its beginning, Searchlight had its own dairy, but the dairy and the farms didn't survive for long. A handful of ranches operated until the early 1990s, when arrangements were made to ban all cattle grazing from the area in order to comply with the federal Endangered Species Act.

Searchlight may have not been favored by nature, but in the years after gold was discovered, this desert place developed into a microcosm of a frontier settlement worthy of historical study.

2

MONEY FROM MASSACHUSETTS

The first accounts of the area around present-day Searchlight came from nearby Summit Springs, which, except for the workings at Eldorado Canyon twenty miles north, was the main center of habitation. The site was believed to be about three miles east of Searchlight, probably at what is now known as Red Well, which is just off the blacktop road to Cottonwood Cove, part of the new Lake Mohave formed after the construction of Davis Dam.

More than a century before the discovery of gold at Searchlight, prospectors combed the entire desert west of the Colorado River for numerous minerals and hard metals, including gold, virtually without success. They found float (loose rocks that when panned showed some value) in some of the washes, but no outcroppings of ore surfaced.

The discovery in Searchlight did not result from this initial investigation. The area had been closely prospected for many years; in Eldorado Canyon mineral exploration had been routinely conducted since the days of Spanish rule. The Colorado River, relatively close to Searchlight, had been freely navigated during the nineteenth century. The intercontinental railroad (the Atchison, Topeka, and Santa Fe) was built only twenty-eight miles to the south, and the U.S. Army and the U.S. mail were moved over the pass near Piute Springs even before the Civil War. So the geography of Searchlight was not unexplored territory.

Some dispute exists as to whether the mining camp that would become Searchlight was discovered in 1896 or 1897. The latter date has been commonly used for almost a hundred years, principally because all federal government publications used it. The pioneers who settled Searchlight and their descendants later disputed that claim and have advocated the earlier date.

It seems clear that Fred Dunn, of Needles, California, about fifty miles south of Searchlight, had for many years corresponded with various eastern capitalists to secure investments in his mining properties. One of those with whom he communicated was a Boston investor named Colonel C. A. Hopkins. In one of Dunn's letters, Hopkins read a description of the Sheep Trail Mine, near Needles. The colonel replied to Dunn, expressing interest in the claim, but by the time the mail was delivered to Dunn, the Sheep Trail Mine was no longer available for purchase.

Dunn again wrote to Hopkins in Boston and told him that although he had been unable to secure an option on the property Hopkins originally desired, other mining claims were available. When he wrote the letter, however, Dunn actually had no properties to offer, so he hired John C. Swickard to locate claims for the consideration of $1 per claim. Swickard began work immediately, concentrating his efforts in the Crescent and present-day Searchlight areas. At that time the Crescent Mountains, ten miles west of Searchlight, were the site of vigorous mining activity because of significant recent discoveries of turquoise. So the general Searchlight area was being investigated with some success before 1896.

When Dunn believed he had enough claims to interest Hopkins, he invited him to come for a visit to inspect the property.[1] Hopkins came to the prospected area but purchased nothing, though he did retain Dunn to look for other properties.

Hopkins exhibited interest in the area around Searchlight because of the preponderance of low-grade ore, which was more than enough to intrigue him. Unfortunately for Hopkins, although Dunn had retained Swickard, the latter owned almost all the property that would eventually make up the claims that became the famous Quartette Mine. The only claims that Swickard did not own were two small fractions of 49.5 feet at either end of the vein that he first saw when he began his work for Dunn. These fractions were claimed by Fred Colton and Gus Moore in 1897. In order to obtain sole ownership of the entire outcropping of the vein, Swickard traded the soon-to-be Duplex mining claim to Colton and Moore in exchange for the fractional claims he wanted.[2]

It seems clear that prospecting in the Searchlight area was inspired not only by Hopkins's investment interest but also the long-standing interest on the part of Dunn, Swickard, and others in the triangle area where Nevada, Arizona, and California met, near the Colorado River. By 1897 successful mineral exploration activities had already been undertaken in the Eldorado Canyon, Goodsprings, and Crescent areas.[3]

Swickard was proud of his Quartette, and the meticulous work he performed for Dunn was evident many years later. His location monuments were unique. A *Searchlight Bulletin* more than ten years after the association carries a description of the monuments, which resembled a pawnbroker's sign consisting of two stones and a pebble.[4] To locate a claim, a prospector would usually put in place a small post and attach a tobacco can to it with the claim notice inside. Because he was being paid $1 for each claim he located, Swickard moved forward in a rapid and wide-ranging fashion, claiming outcropping after outcropping.

Swickard decorated the Quartette property with large signs that carried this message: "Any sheepherding sons of bitches that I catch digging in these here claims I will work buttonholes in their pock-marked skins."[5] Since Swickard was always heavily armed, his threats were heeded.[6]

Even though Swickard was extremely protective of his claims, he shortly sold them to the trio of Benjamin Macready, a Mr. Hubbard, and C. C. Fisher for a team of mules, camping equipment, and $1,100. Though proud of his effort in locating the Quartette claim, he sold because he had no faith in the property; he believed the outcroppings were a blowout of the vein and would have no depth. By today's standards the consideration he received for his claim seems paltry, but by the standards of 1898 and 1899 the payoff was significant. It had been known since 1896 that low-grade ore existed in the area that became Searchlight, yet no exploration of more than a hundred feet in depth had taken place, not even by 1899, when Macready sold the Quartette to Hopkins. There is some evidence that Macready obtained the interests of Hubbard and Fisher and then combined his holdings with Dunn's before selling to Hopkins and Associates.[7] The selling price this time was $150. Before Hopkins could accept the deal, the price was raised to $200. Highly insulted, Hopkins felt he should not consider the new price. His mining engineer, Leo Wilson, intervened and for an additional $50 Hopkins increased his fortune.[8]

Dunn and Macready were forced to sell the Quartette property because they had been unable to raise the capital for an ongoing mining operation. After the sale, however, they remained involved in the new operation. Dunn served as the resident agent of the corporation, and Macready acted as Hopkins's superintendent. Each maintained a minor ownership, but the real financial force was the Bostonian, Colonel Hopkins.

Money from Massachusetts had a similar impact on another mining venture, in 1904, in the Robinson mining district of White Pine County, Nevada. James Phillips Jr., a New York financier, and Mark Requa, one of the owners of claims in the Comstock Lode, persuaded the Loring brothers of Boston to capitalize the Nevada Consolidated Copper Company, which later led to Kennecott's massive copper mine and processing facilities near Ely.[9] Some say that without Massachusetts money, that important Nevada operation could never have been developed. In fact, a look back through history shows that nearly all of Nevada's mining enterprises were funded from outside the state, except for a few operations developed later in the century by Nevadans like George Wingfield.

3

THE CAMP THAT DIDN'T FAIL

One of the real difficulties facing early prospectors in southern Nevada was that to file a claim, they had to travel more than 200 miles to Pioche, a trip that took at least ten days. This presented great hardship, especially in the winter months, when the weather conditions around Pioche could be severely inclement.

As early as 1898, articles appeared in periodicals touting the discoveries made in the Searchlight area. The references were actually to Summit Springs, with directions to the specific site, for Searchlight had not yet been named.[1] The most definitive citation observed the following: "At this point, fifty miles north of Needles, California and some ten miles west of the Colorado River, there is some excitement caused by a promising gold strike made by a Mr. Colton. His first shipment of the selected ore yielded at the rate of 72 ounces per ton. He is now shipping a carload that is expected to produce some 200 dollars per ton. Conservative miners who have recently visited the locality are pleased with the outlook in this vicinity."[2]

On July 20, 1898, the mining district of Searchlight was formed. The place chosen for the undertaking was the only frame or wooden building in the whole camp,[3] a little shack located near the present-day Cyrus Noble Mine, not far from where the Santa Fe Railway depot would later be situated. The founders were described nine years later as a "small bunch of adventuresome spirits who had undertaken the task of unbuckling the girdling of the gold that encompasses this immediate mineralized section, and [took] advantage of the privileges allowed them under the United States mining laws."[4]

The group of miners and prospectors involved in forming the district drew up a set of bylaws and regulations. Rather than drafting a list of crude, misspelled rules, they put into effect a concise, systematic, and businesslike set of standards covering every point necessary for the filing of a mine claim.[5]

The formation of the mining district did not obviate the need for the ultimate filing with the county recorder in Pioche, the seat of Lincoln County. Because Pioche was so far away and winter weather often made travel impossible, principals were allowed to establish the priority of the claim by filing it initially with the district recorder, then transfer the documentation to Pioche at a convenient time. This arrangement prevented many claim disputes. The original papers of formation were written on ordinary notebook paper in handwriting and then pasted in a rusty book, which, as of July 19, 1907, was still preserved in the recorder's office.[6]

Those who signed the formative papers were E. J. Coleman, who acted as chairman; G. F. Colton, who acted as recorder; Samuel Foreman; S. Baker; F. C. Perew; F. W. Dunn; H. P. Livingston; C. C. Fisher; T. B. Bassett; J. F. Dellitt; W. O. Camp; W. G. Lewis; G. B. Smith; and E. R. Bowman. It is interesting that the two accounts of the formation of the district agree on everything except one of the signatories of the handwritten document establishing the mining district. The *Searchlight*

Bulletin of July 14, 1911, lists a woman by the name of Mrs. Hattie Cook as one of the signers, but an earlier account in the same paper on July 19, 1907, does not mention her name. It may have been merely an oversight that the name of the only woman who signed was left out, or it might have been a subtle denial of a woman's role in the founding of the town. Hattie Cook did, however, subsequently locate her own mining claim, the Flat Iron.

Many claims had been recorded in Pioche before the formation of the Searchlight district, including Fred Colton's initial discovery, which started the rush to the Searchlight area. But the first claim actually recorded as "Searchlight," called the Happy Jack, was located on May 3, 1898, just a few days before the formation of the new Searchlight district. This initial claim was located by J. F. Dellitt, one of the people who formed the district.[7] The discovery of the big claim by G. F. (Fred) Colton on May 6, 1897, was not actually recorded in Pioche until the next January. From this example alone it is clear why it was necessary to form the district.[8]

By October the mining camp had its own post office. That same winter many more claims were filed, with the accompanying speculation that all of them would yield riches. These reports sparked an increase in the flow of people to the new camp.[9]

The development of Searchlight came at an opportune time in the history of Nevada, since the Comstock Lode was all but exhausted by the time Colton struck gold in 1897. The shipment of ore from the Searchlight district followed a twenty-year slump in Nevada mining and gave the state increased visibility nationwide.[10]

It later became apparent that any ore of significant value in Searchlight would be found at depths of more than 200 feet. Extracting ore at that depth was usually prohibitively expensive for individual prospectors; consequently, many operations followed the example of the Quartette and consolidated their efforts.

The *Engineering and Mining Journal* often reported on such consolidations. Among the transactions recorded there was the New Era Mining Company, which incorporated in 1900 with $300,000 in capital, a significantly large amount of money at the time. The Duplex claim was developed with financing out of Riverside, California, allowing the construction of a mill and extensive underground development. The Searchlight Mining and Milling Company, known thereafter as the M&M, was capitalized in 1899 with sufficient financial resources for continuous work until ore was finally found in 1904.[11]

But the Quartette was the mine that propelled Searchlight out of the ranks of insignificant Nevada mining towns. The Quartette was a great mine by any standard, and its dramatic success allowed Searchlight to become a mining camp of world-class proportions.

The finest mine in Searchlight almost never came into existence, however. The original capitalization by the Hopkins group was soon expended, but more money was sunk into developing the mine. Suddenly, Fred Dunn, the company's resident agent, acting on instructions from the owners in Massachusetts, ordered the foreman, Jack Russell, to stop work. Russell politely but firmly informed Dunn that he took orders only from superintendent Macready, who was in Los Angeles. Dunn then contacted Macready in Los Angeles by telegraph, ordering him to close down the mine. Macready could not return to Searchlight for four days, since the train from Goffs to Manvel ran only three days a week. He did not receive the message from Dunn until Thursday, so he had to wait for the Monday train. Instead of biding his time until the train ran, Macready wired two words to his foreman: "Crosscut south,"[12] instructing the men to continue work but to extend the work at an angle rather than straight down.

When Hopkins originally purchased the Quartette, the shaft was 100 feet deep.[13] At the time of the apparent depletion of funding, the shaft had reached the 300-foot level, and the findings were not encouraging. In fact, the ore was averaging only $3.84 in gold per ton. Since all of the ore in other Searchlight mines was being found at depths of less than 100 feet, Benjamin Macready was actually charting unknown territory when he ignored the instructions from his owners and ordered the miners to continue. When he arrived in the camp four days later, they had struck a bonanza—and they had reached the ore after only two more shifts.[14] By the time the mining boom ended, the Quartette accounted for more than 50 percent of all the gold taken out of the Searchlight mining district.[15]

Twenty-three miles southwest of Searchlight was a railroad connection, originally called Barnwell after the first telegraph operator at the station. The Quartette and other Searchlight operations had to haul ore over this twenty-three miles of incredibly rough terrain in freight wagons to the small railroad line, originally called the Nevada Southern and then the California Eastern. From here, the ore was shipped to the central complex of smelters and mills in Needles, California. It was a time-consuming and expensive operation. To curry favor with the Atchison, Topeka, and Santa Fe, to which this thirty-mile line connected, in 1893 the small railroad changed the name of Barnwell to Manvel, for the Santa Fe president. The small railroad was taken over by the Atchison, Topeka, and Santa Fe in 1901. Shortly after Searchlight was discovered, the president died, and the name of the site was changed back to Barnwell. Ultimately, the railroad built a line to Searchlight.

With the significant gold production at the Quartette, and the long, hard haul to Barnwell, management agreed to finance the construction of a mill at the Colorado River, about fourteen miles east of Searchlight. The haul to the river made sense because the load would be heavy going downhill and the freight cars would be empty on the arduous trek back up the hill. The construction of the mill at the river also solved the problem of the lack of water in the immediate Searchlight area. In fact, even at the 300-foot level, where the big strike had occurred, there was no sign of water, at the Quartette or at any other place in the camp.[16]

Building the mill was not a difficult engineering task, but constructing a railroad to the river was more complex and expensive. It was, however, necessary in order to save costs in the production and processing of the ore, and so the decision was made to proceed. The construction of the mill and narrow-gauge railroad took nearly a full year, until May 1902. The mill ran continuously until June of the following year.

A significant water supply was finally reached at the Quartette about the 500-foot level, at just about the time when the mill and railroad construction was completed.[17] The discovery of water in the mine reduced the need for the riverside mill.

4

WHY SEARCHLIGHT?

Theories about how the town of Searchlight was named have provided ongoing controversy among the area's residents almost since the founding of the town. One of the first mentions of the name Searchlight occurred in a mining journal of February 11, 1899: "Miners flocking to the Searchlight camp located about 100 miles north of Needles. Highgrade gold quartz veins have been discovered."[1] Note that the specification of Searchlight's location is off by almost fifty miles—Searchlight is only fifty miles from Needles, not a hundred miles.

After Colton's initial discovery, the exploration and mining activity began in earnest. It is noteworthy that even though Colton and his family lived in Searchlight throughout most of the next fifteen years, with brief visits to California, neither he nor the family commented on the initial prospected discovery. No interviews with George Frederick Colton, the founder of Searchlight, can be located in which he explains the details of his location of the Duplex, or even how the name Searchlight was assigned. Several competing versions of the town's naming have been proffered, and Colton neither confirmed nor objected to those differing versions. For example, descriptions of how the camp got its name appeared in early Searchlight newspapers at a time when Colton was a prominent citizen of the town. In the decades following the decline of Searchlight, he came in and out of the town, and members of his family lived in nearby Las Vegas, but he left no traceable interview in which he discusses the naming of the camp. During this time, however, other theories emerged about the naming of Searchlight.

One version insists that it was named for an early miner in the area, Lloyd Searchlight. There is, however, no record of anyone by that name who ever lived nearby. The confusion developed when a man known only as Mr. Lloyd started the Lloyd-Searchlight Mining Company, a company that didn't begin operations until the Searchlight mining district had long been formed and named.

A *Bulletin* headline in 1906 read, SANTA BARBARANS PAY $40,000 FOR BONANZA PROSPECT. LUCKY OWNERS RETAIN LARGE INTEREST—WILL BE KNOWN AS LLOYD-SEARCHLIGHT. The article goes on to state that the development work would be under the direction of Mr. Lloyd. "Although the local management is preserving clam-like silence, it is learned on the best of authority that the Lloyd-Searchlight has struck it rich. In point of discovery and development Lloyd-Searchlight is the foremost property at Camp Thurman, fifteen miles south. Its owners all reside in Santa Barbara, California."[2]

A second version is more humorous. Prospectors congregating at Summit Springs before the formation of the Searchlight district used to joke about the miners John Swickard and Joe Boland, who patiently ground their very low-grade ore in a mule-drive crusher, saying, "There is ore there alright, but it would take a searchlight to find it." It was recalled that they all laughed afterward, but when Fred Colton turned up some high-grade ore three miles west of Summit Springs, he remembered this joke and called the location Searchlight.[3]

A *Searchlight* newspaper article lends credence to this version because Colton and various members of his family were living in Searchlight when it was written. Logically, if the story were inaccurate Colton would have denied it. Conversely, it could also be argued that if the story were not true, Colton would not want to contradict it, since the tale gave him greater standing in the town.

The newspaper stated in 1906: "It might be interesting here to relate how the camp originally got its name. A number of prospectors had discovered some float in the valleys to the east and west of town and had a camp established in a gulch near where the Cyrus Noble is now located. Coming into camp one evening tired, sore and disgruntled, Fred Colton, the first discoverer of the camp, threw his canteen on the ground and exclaimed, 'there is something here boys, but it would take a searchlight to find it.' Two or three days later he found the ledge of the present Duplex and named it Searchlight. And this was the christening of the camp."[4]

Another recently unearthed version of the town's naming was buried in a 1911 *Bulletin* article. In naming the mine and the town, Fred Colton was impressed with the wonderful view from the Duplex Mine, which was situated on a large hill overlooking the town. He is reported to have said, "This would be a nice place to mount a searchlight."[5]

Yet another version of the unusual name Searchlight originated with a box of wooden matches, which were essential for lighting cigarettes, cigars, stoves, and for general survival in the early part of this century. One of the most popular brands was named Searchlight. The story is told that a handy box of Searchlight matches was seen at the camp and inspired miners to give the name of Searchlight to the desert mining district.[6]

George Colton's grandson, Gordon, has perpetuated the matchbox version of the tale, spreading word that this is how the town got its name. Gordon was very loquacious, but he did not base his story on conversations with his grandfather. He never lived in Searchlight until late in his life, and the box-of-matches version of the story didn't appear until many years after the camp was founded. (As an interesting side note, Gordon was alleged to have played five years of high school football at Las Vegas High School before embarking on a professional football career with the Los Angeles Rams. His classmates even assert that he was All-State at two different positions. This is confirmed by his son, Stanton, a former Nevada state treasurer. In his old age, Gordon became the constable and deputy coroner of Searchlight.)

Most longtime residents of Searchlight agree that the name came from Colton's being told—or saying—that one would need a searchlight to find gold, but there is no surviving interview at any time with the original developers of the mining district that would shed light on the authenticity of this version. The Searchlight newspaper opined, however, that this version had "the widest credence."[7]

The most credible version of how Searchlight got its name is Colton's story of the need for a searchlight to find the ore. A few have felt more support for Gordon Colton's box-of-matches theory. But before a jury both his story and the other versions would fail. Historian John M. Townley agrees, conceding that the most logical version of the name's origin is the one that centers around needing a searchlight to find the gold, even though Colton never commented on the naming.[8] Townley does confirm, however, that even five years after the discovery of gold in Searchlight, no one was certain as to the origination of the name.

George F. Colton, the town's founder, was rarely interviewed on any subject having to do with the beginnings of the town. In 1906 he returned after about a year's absence from Searchlight and said, "I came here in 1897 and pitched my tent near the present site of the Searchlight Hotel. . . . This is not only a camp without a failure, but a camp with a future."[9] Jane Overy, resident historian of Searchlight and the curator of its museum, insists that the place where Colton pitched his tent is the present location of the post office parking lot.

Possibly another reason that Colton did not make a big deal out of his discovering the town and naming it is that perhaps in his mind he didn't do either. It is clear that there was significant prospecting in the area of Searchlight long before 1897. This is confirmed by many sources, not the least of which is a news article describing Colton as "the father of Searchlight, because of the fact that he discovered the first interesting claims in the camp and built the first house in town."[10] The article recognizes him not as the discoverer of gold in Searchlight but as the discoverer of the first interesting claim. George Colton died in California in 1916.

The same newspaper rightfully calls John Swickard the father of Searchlight. In the early nineties, Swickard prospected through this territory. His locations were the first in the district. As early as 1896 he enlisted the backing of Colonel Fred Dunn and, in 1897, he established the first permanent camp, at Hall's Well.[11]

In short, Searchlight is a camp with more than one father.

5

THE BIG STRIKE

The purchase of the Quartette by the Hopkins group was important to the success of Searchlight. Without the large initial infusion of capital into the Quartette operation, the mine would not have been sunk deeper than any other mine in the history of Searchlight. Without the deep shaft and the subsequent huge ore strike, mining in this area would never have developed. The extensive mining and exploration that later occurred was all based on the early success of the Quartette.

The *Searchlight,* the newspaper of early Searchlight, promoted the town as the "camp without a failure." Until 1907, when the newspaper changed its name, this phrase was on the masthead, proudly broadcasting the area's prosperity to the state and nation. The newspaper hoped to attract new capital and people to the southern part of Nevada.

Shortly after the fateful telegram was sent by Macready, the Quartette seemed destined to become a real bonanza. By 1903 Searchlight was the talk not only of Nevada but, according to the local newspaper, of the whole mining world. At what seemed to be the height of Searchlight's success, however, labor problems erupted.

Union activity in Searchlight was the result of organizational efforts of the Western Federation of Miners (WFM), founded at Butte, Montana, in 1893, shortly before the discovery of gold in Searchlight. Some have written that the creation of this union was the "birth sign of the coming militant industrialism of the Industrial Workers of the World. In the first decade of the twentieth century this union enjoyed success in its activities in Goldfield and to a lesser extent in Tonopah."[1]

A costly labor strike almost brought the mining boom in Searchlight to a standstill in 1903. Even though the union focused on the Quartette, other operations panicked, and most closed down until the strike was resolved.

The union strike, called on June 1, 1903, was precipitated by a number of disputes, primarily a law passed on February 23, 1903, by the Nevada Legislature that limited the workday to eight hours for most mining-related jobs, particularly underground positions. On June 1, the mine owners posted a notice ordering all workers not affected by the new law to work nine hours. This gave the union an issue. It is interesting to note, however, that the only workers affected by the Quartette order were three men who did not work in the mine or the smelter but were hoistmen and trammers who were not covered by the new law. The law stipulated an eight-hour workday not only for underground miners but also for those who worked in smelters and all other positions involving the reduction of refining ores and metals. The strike was ostensibly called because of these three men, but it also provided an opportunity for this new labor organization to flex its muscles.

Initially the union had the support of most people in town, who thought that the workers deserved better pay and improvements in working conditions. Even the newspapers that covered Searchlight, the *Searchlight* and the *DeLamar Lode,* appeared to favor the goals of the striking workers. It also was clear that the real issue was not the basic economic one but whether labor or management was to control the Searchlight workforce.

After the company's notice was posted, the union committee asked the mine superintendent what would happen if the union called the workers off the job. Management replied that the mine would be closed.[2] In fact, the mine was closed on June 1, without the union's ordering a work stoppage.

Anticipating union action, the owners of the Good Hope Mine and the Duplex also terminated operations the day after the shutdown by the Quartette owners. The provocative nature of the mine owners' actions is clear when one realizes that in all three of these mines, only three men were working nine-hour shifts, and that was at the Quartette.

At first, public comment about the way the union was conducting the strike was very positive. The press and Searchlight residents were favorably impressed that there was no violence. In fact, the union helped foster positive public relations by allowing four of its union men to be engaged in working the pumps at the lower levels in the Quartette, where water would have accumulated, damaging equipment and the workings in the shaft and drifts, if the pumps had not been kept operating.

The union movement in the western United States was in its infancy at this time, especially in the mining industry. Strategies for resolving impasses between labor and management were not well developed, and the two parties were experimenting with ways to end disputes like this. The union wanted to appear tough and strong, even resilient, and the mine owners wanted to put an end to the union before it gathered strength.

In the early days of the strike there was considerable talk of arbitration, but that was very short-lived. In the *DeLamar Lode* of June 23, the prospects for settlement were more vague than ever. In fact, the mine owners and managers had left for Los Angeles almost as soon as the strike started. The owners indicated they would receive union representatives only in Los Angeles, stipulating that all negotiations would have to be conducted somewhere other than in Searchlight. This action only made relations between the warring parties worse, since during the early days of the strike various union representatives from the national office often visited Searchlight with the intent of negotiating with the owners. They soon learned there was no management to meet with unless they went to Los Angeles. The papers reported: "J. H. Vaughan, representative of the miners union, was in the city Monday to see if the mine owners had anything to say, or to see if they were desirous of a conference."[3] The same newspaper article observed: "John C. Williams, Vice President of the Western Federation of Miners, is expected to be in camp tonight to take hold of the union and end the strike."[4]

The local newspaper strongly condemned the owners' and managers' retreat to California at the strike's inception. Precisely, the *Searchlight* also reported in its June 26 edition that it was the employers' intention to create an issue to discredit the union. Again, the paper and the townspeople clearly were on the side of the miners and not the owners.

Because unionism was new in Nevada, and this type of labor unrest was fresh in the western states, the union representatives were continually trying to justify their ability to sustain a long strike. When the owners, in effect, refused to negotiate, the union announced that it had ample funds to support the union miners for an indefinite period of time. The company responded by announcing a policy inviting nonunion men to apply for jobs.[5]

As the contention continued, so did the competition for the most marketable story describing the strike. Since the *Searchlight* was published in Searchlight and was the paper closest to the controversy, it seemed always to paint a picture of peace and serenity during this time, wanting only to project the image of a boomtown. The *DeLamar Lode* was the newspaper in the town of Delamar, located in what was then the upper part of Lincoln County, about 30 miles from Caliente and 150 miles from Searchlight. On August 4, only two months after the beginning of the dispute, the *Lode* opined that only the bad people in the county were left in Searchlight. The *Lyon County Times,* published in Yerington, about 350 miles north of Searchlight, reported that the miners at the Quartette struck to have their workday reduced from twelve hours to eight. Such a report was ridiculous; at no time in Searchlight's brief history had the miners been required to work more than nine hours. The *Searchlight* wrote a rare scathing editorial, attacking those who engaged in perpetuating false rumors and emphasizing that the strike was being conducted in a peaceful, orderly manner, on issues that were strictly a matter of principle.

On July 3, 1903, Judge M. A. Murphy of the state district court struck down the law establishing the eight-hour day in mining-related work, ruling it unconstitutional. The court declared that the legislation, being class in nature, was invalid because it separated mining and milling from other types of employment, violating the state constitution by taking property without due process. In effect, the court ruled that the Nevada Legislature had no right to dictate hours to miners and mill workers when it did not set the same standards for other types of work. Because of this, the owners were being forced to work their property under conditions that unfairly paralyzed them by having their employees work fewer hours than other workers.

Immediately thereafter, the union, through John Williams, a vice president, approved the strike despite the court's interpretation of the recently passed law. The union again declared its ability to withstand a long strike, since the WFM had supplied the funding necessary for the duration of the union activities.[6]

Even though the owners and managers were not often in the vicinity of Searchlight, they had obviously been plotting to ruin the union and end the strike. Their first move was to form the Desert Mine Operators Association. Although the association's bylaws prohibited discrimination against union members, everyone knew that the purpose of the organization was to stamp out the union. The association even included mines in California, as well as Searchlight's Quartette, Duplex, Good Hope, New Era, Cyrus Noble, Southern Nevada, and Ranioler. The formation of this association was the beginning of the end of the effectiveness of the labor movement in Searchlight. The owners began to investigate ways to reopen the mines with or without the union miners.

The commercial interests in town formed a citizens' committee to arrange a conference with the owners and the union and to act as a liaison, carrying messages of hope between the two warring parties. The Quartette officials, representing all the other companies, refused to talk to the union but professed a willingness to resume operations and to take back all former employees—with the same wages and hours that were in effect before the strike. Company officials also indicated that when the mines began making a profit again they would entertain a different wage scale. This decision by the owners meant that underground workers, as well as blacksmiths and engineers (who were traditionally treated like miners), would work eight hours and aboveground men would work nine hours. All others, such as laborers and those on temporary jobs, would work nine-hour shifts. The union rejected that offer, holding out for a fifty-cent raise and an eight-hour day for all mine-related work.[7]

It didn't take long for businesses to start feeling the effects of the mines' closure. Though there was significant independent prospecting being conducted during the labor unrest, it generated very little commercial trade.

The first strikebreakers, two miners from Los Angeles, arrived in September. They didn't stay long, since they were persuaded by the union not to go to work. Several days later, two stagecoaches arrived with men who were to begin work at the Quartette. The *Searchlight* of September 25 reported that the Quartette had gone back into operation with thirty-five men on its payroll, including miners and guards. Even though this is a small number of employees, the company's action demonstrated its determination to get the valuable property back into production. Conversely, the union was doing everything it could to prevent the mine from adding employees, even stationing pickets at various locations, like Manvel, Ibex, Needles, Goffs, and San Bernardino, to deter the further importation of strikebreakers and other anti-union activities. The union also appealed to other labor organizations in Los Angeles and San Francisco, urging them to make every effort to keep workers from coming to Searchlight until the strike was settled, and it advertised in the Joplin, Missouri, area—the home base of the union—to warn hirelings of the situation in Searchlight.

By early October, however, the Quartette had started a stamp mill, located next to the mine. It was obvious to all that for the mill to operate the mine had to be producing ore. Nevertheless, the union still talked as if it was winning the dispute, even though it was apparent that the mine was operating with nonunion workers.

One incident that added to the excitement during these tense times was when the union learned that thirty strikebreakers were en route by train to Manvel, on their way to the Searchlight mines. The union organized a march along the twenty-three miles from Searchlight to Manvel to intercept them. After the long, grueling walk, however, they learned that not a single strikebreaker was on the train.

The editorial position of the *Searchlight* took its first turn against the union on October 2, noting that the union was hurting its own cause by not working harder to resolve the dispute. Recognizing that nonunion men were already being shipped in to work, the editor further elaborated that the new law, on the basis of which the strike had been called, had since been declared illegal. The article made the case that the two sides were crushing the life out of the new town of Searchlight and stated that the business of the town was being ruined and the storekeepers forced to operate at a loss. This was the first editorial calling for an end to the strike.

Just a week after this editorial appeared, the Quartette, the Good Hope, and the Southern Nevada mines were back to full operation. Simultaneously, the union suffered several other setbacks, including the arrival of twenty-one workers from Joplin, Missouri, and several more from the mines of Colorado. The strikebreakers went to work under the conditions that had existed before the strike began.

About this time, the Quartette opened its own general store and even built bunkhouses for its workers, which provoked extremely negative reactions from both the merchants and the general population. The Quartette, located about a mile and a half from the center of the city, was becoming its own town.

The *Searchlight* condemned the actions of the mine owners. They were particularly galling to the paper because it had recently run editorials supporting these companies. In desperation the paper called on the union to end the strike, but the union remained defiant. The newspaper finally declared that the union had lost the goodwill and sympathy of the community.[8]

Even at its most intense, however, the strike in Searchlight was orderly and nonviolent. The sheriff from the county seat of Pioche periodically visited Searchlight to monitor the situation, always returning with reports of nothing more than rumors of disturbances. The entire period of the strike was unusually calm.

The peaceful nature of the Searchlight strike was similar to the minimal labor strife that the Comstock had experienced a generation earlier. The Western Federation of Miners formed its first union in southern Nevada in Tonopah in the summer of 1901, and Tonopah escaped any real labor problems until after World War I.

At nearby Goldfield, however, there were significant labor disputes, marked by numerous episodes of violence. The unions had obviously learned from the losses of the WFM in Searchlight, for they became powerful in Tonopah and Goldfield. In 1907 Goldfield was an armed camp. Several shootings occurred, with one reported death. Eventually President Theodore Roosevelt, at Governor John Sparks's request, sent federal troops to quell the quarreling factions. In comparison, Searchlight had been very calm.

In January 1904 the courts again surprised the entire Nevada mining community with a long overdue decision. The Nevada Supreme Court overruled the district court and declared the wages-and-hours law constitutional. The reason for the strike had come full circle. But like its predecessor, this final decision did not change the fact that the union had been broken. The union continued operating in a strike mode for the next year, even though almost all of the union men had gone back to work. Those who returned to the mines were required to sign a card agreeing not to participate in union activities, pursuant to the Desert Mine Operators Association rules.

In 1907 the same card system was put into place in Goldfield just before Roosevelt ordered federal troops to Esmeralda County. The right of the companies to have employees sign such a card was affirmed by the Nevada state legislature in 1907.[9]

The strike had a tremendous impact on the new town. The merchants suffered in not being able to develop commercial enterprises as quickly as they otherwise could have. Many people experienced economic hardship as a result of the strike, and workers with known union sympathies were laid off. For example, James Lappin, foreman of the Quartette Mine, was laid off as a result of his union leanings. His wife, Lula, opened an ice cream parlor to provide income for the family, but the store failed and the Lappins migrated to Southern California where, at age fifty, James began a second career as a farmer. He died in Anaheim, in 1908, at age fifty-five, just about five years after being run out of Searchlight.[10] James Lappin's story was repeated numerous times in the lives of the early inhabitants of this boomtown.

The labor-management problem in early Searchlight had a very limited effect, however, setting the progress of the town back for only about three months. Though the union and townspeople kept referring to the "strike," in reality it didn't exist—the strike was actually broken early in the dispute.

As mentioned above, much of the friction was caused by the competing newspapers, the *DeLamar Lode* and the *Searchlight*. The *Lode,* for example, had the strike settled by September 20 when it reported: "The backbone of the strike is broken. The Quartette landed a number of men on its property yesterday to begin work and to date things were moving as of old." The *Searchlight* was more cautious. In its October 2 issue, it reported: "The strike situation this past week has shown little change."

The strike did have other, unintended consequences, however. It was because of the pro-union stance taken by the *Searchlight* and some of its advertisers that the Quartette Company decided to start its own general store and other competing businesses at the mine site. The action was clearly an attempt to punish those businesses that went along with the union leaders.

The dispute also caused the company to focus on labor relations instead of on ways to improve the mine. One of the Quartette's managers said in December 1903 that if the strike had not occurred the company would have built a railroad from Ibex to the camp.[11]

In just three short months, the union was vanquished. Though the exact date of the defeat is debatable, the conclusion is not. Union activity disappeared and to this day has never reappeared in Searchlight.

6

THE BIG MINE

If one travels to Searchlight today and drives or walks around the area, he or she will see scores of mines, mine dumps, tailing remnants, gallows frames, and even collapsed mill sites. The names of the mines are entertaining and curious: Empire, Good Hope, Good Enough, New Era, Blossom, Key, Tiger, Barney Riley, Rajah, Yucca, Shoshone, Ironclad, Parallel, Searchlight Mining and Milling (M&M), Western, Berdie, Pan American, Elvira, Mesa, Pompeii, Southern Nevada, Telluride, Empire, Red Bird, Blue Bird, Saturn, Santa Fe, Philadelphia, Eddie, Ora Flame, Carrie Nation, Magnolia, Hyacinth, Poppy, Parrot, Spokane, Cushman, Dubuque, Golden Garter, Silk Stocking, Eclipse, June Bug, Little Bug, Cushman, Duplex, Water Spout, Cyrus Noble, Golden Rod, Water Wagon, Bellevue, Chief of the Hills, Crown

King, Quaker Girl, Iditarod, Greyhound, New York, Stratford, Quintette, Columbia, Gold Legion, Calivada, Annette, Gold Coin, Gold Dyke—these are but a sampling of the myriad claims that make up the Searchlight mining district. A few of the mines were sporadically good producers, especially the Duplex, Blossom, Good Hope, and Good Enough. The remainder produced little in the way of substantial ore. Most of the shafts were inclined; only a few were vertical. Most were shallow, not delving more than 200 feet in depth. The Duplex, Pompeii, and Good Hope were unusual in that they were sunk more than 500 feet deep.

It was, however, with anticipation and great hope that the early Searchlighters approached the future. In May 1904 the headline in the local newspaper blared that the area was THE PREMIER DESERT MINING DISTRICT.[1] The first years of the boom created much speculation and investment. By 1904 there were seven mills within a mile of one another: Cyrus Noble, Quartette, Duplex, Southern Nevada, Good Hope, M&M, and Santa Fe.[2] Unfortunately, soon after construction, several of the mills were left without any ore to process.

The Cyrus Noble earned its name because the claim was sold for a bottle of Cyrus Noble whiskey. Ten days before the assessment work on the claim was due, the owner walked into a Searchlight bar and shouted, "What am I offered for my claim?" "I'll give a cigar," one patron said. The offer was accepted. Immediately afterward, the new owner crowed, "What am I offered for my claim?" Another miner responded, "I'll give you this bottle of Cyrus Noble." "Sold," replied the new owner. The third owner made a good bargain because, unlike many others, this claim did produce some gold.[3] Adjacent to the Cyrus Noble were other claims with names that related to the bottle, such as the Little Brown Jug.

The Duplex was the second-best mine in Searchlight, but it was a very distant second place. Another good mine was the Blossom, which was staked by George Butts. It produced a small amount of high-grade ore, but Butts didn't have the money to work it. While trying to sell the mine, he lived on the property in abject poverty in a hut built of Joshua trees. For more than a year he lived in these harsh conditions, holding out for his price. George Butts was given many offers for the claim, but he held

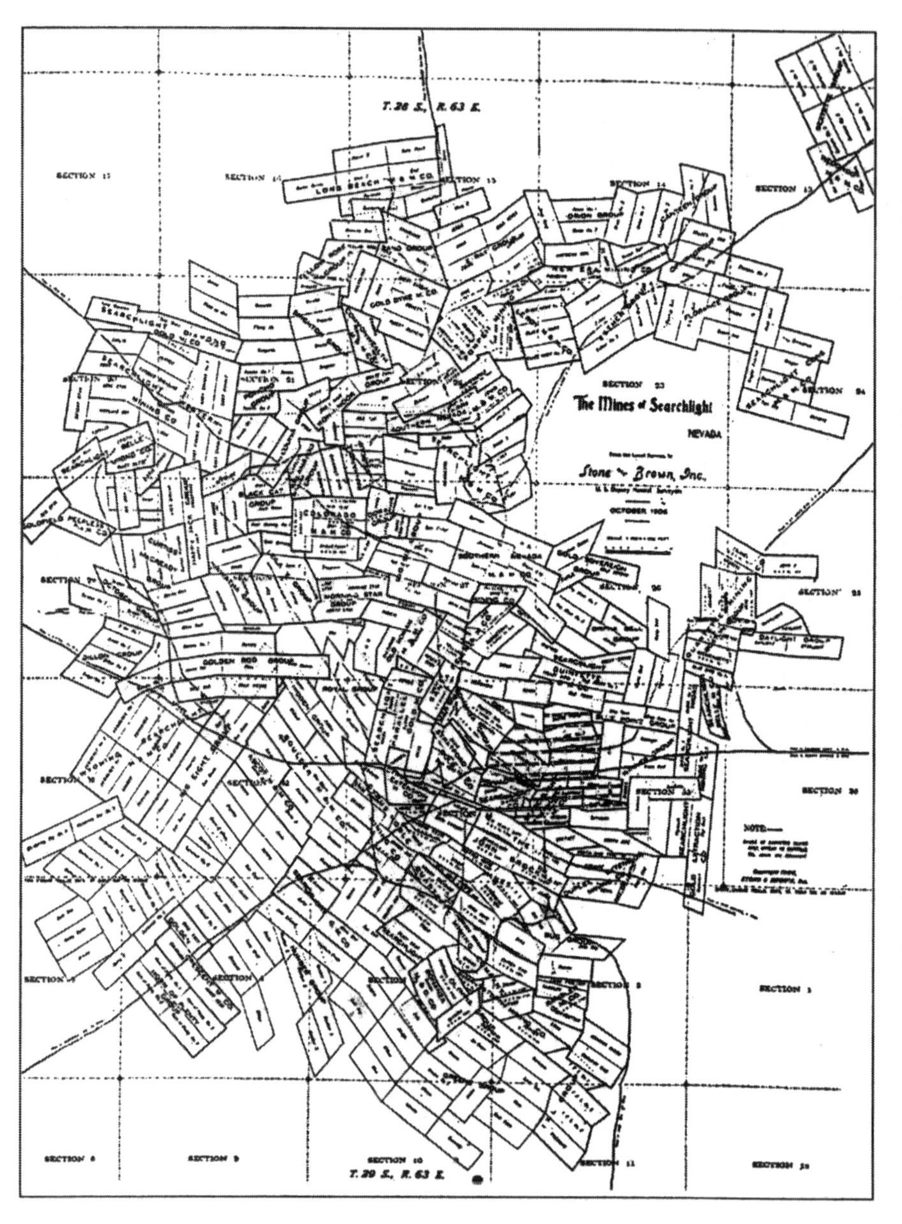

Map of Searchlight Mining District. Courtesy of Stone & Brown, Inc.

out for $25,000, a huge price in that day. After almost two years had gone by, he got his $25,000. He died three days after the sale.[4]

Speculation was not limited to minerals. In December 1907 news reached Searchlight that oil had been struck midway between the town and Needles. According to the story, the Wayne Oil Company was confident that a large oil deposit lay beneath the surface. Like many other strike rumors, this one also went bust. The story was never mentioned again, but the anticipation must have been intense.[5]

The only real world-class mine in the history of Searchlight was the Quartette. From 1899, when Macready disobeyed the order to stop further work in the Quartette, this mine became Searchlight's biggest and best.[6] For the first decade of Searchlight's existence, the Quartette was the premier mine. Anyone writing or talking about the camp lifted up the mighty Quartette as a beacon of Searchlight's progress. Even after mining had all but disappeared in the area, it was still a fine mine, continuing to produce small amounts of gold up until the 1960s. It was the best in Searchlight.

During the decade of mining dominance, from 1899 to 1908, not only was the Quartette the biggest producer in the whole of southern Nevada, but several times it was also the largest producer in Nevada and one of the biggest in the entire United States.

From its inauspicious beginning, the Quartette developed into a mine with multiple shafts. The main shaft, or the glory hole, was sunk to a depth of 1,350 feet. As with many mines of the day, an air shaft was usually sunk to help with the circulation of air in the main shaft, its drifts, crosscuts, and other diggings. The Quartette was no different; it used an air shaft that initially started at the 600-foot level and then was raised to the surface. Eventually the shaft was extended down to the 900-foot level when bad air necessitated that fresh air be circulated to the lower levels. Other shafts sunk over the years were distinguished by the names the Carlton, the Crocker, and Shaft #3. These were not cut to great depths, and most were used for ore exploration purposes.

W. J. Sinclair, one of the first dozen men to enter Tonopah and one of the wealthiest men of Nevada, stated in 1904: "I doff my hat to Searchlight, for you certainly have in the Quartette, the biggest gold mine in the country. I have seen many wonderful showings, but never the equal of the Quartette."[7] The *Searchlight* newspaper opined shortly thereafter: "Searchlight is justly proud of the Quartette mine for it is, as it stands today, the biggest and best mine in the Southwest. As a free milling proposition it is unequalled by any mine in the United States, and considering the amount of development done it is one of the largest in the world."[8]

There was a strong basis for this optimism. In November 1903 the mine was working three full shifts, and the mill would begin working three shifts by early 1905. Modern equipment was installed that allowed electric arc lights to shine in the night desert sky, pointing out the location of the famous hole in the ground. The electric lights on the surface were duplicated in the underground workings as well. In 1904 the Quartette milling operations were electrified. There were telephones on the surface and in certain stations underground.[9] At no time, however, did the Quartette Company share its electrical power generation capabilities with the town. Searchlight would later have to develop its own system of electricity.

After the cessation of mining activities in the mine, there was still much talk of the width and depth of the Quartette ore vein; it was indeed the stuff of which legends are made. At the 700-foot level the ore body was described as being more than 40 feet high and averaging $100 per ton, a figure representing more than four ounces per ton. By today's standard this gold would be worth more than $1,500 per ton. Currently, gold ore in Nevada is mined at significantly less than four ounces per ton; many mines are worked when the ore grade has only one tenth of an ounce per ton and sometimes even less. At just one station at the 700-foot level, the stope (a steplike excavation underground for the removal of ore that is formed as the ore is mined in successive layers) was described as being 18 feet by 40 feet and needing 18,000 square feet of timbers for just that one station.[10] By 1906, when the mine had reached the 900-foot level, the vein was measured to be 60 feet wide.[11] In addition to these huge bodies of moderately good ore, another strike occurred on the 700-foot level, which assayed an astounding forty-four ounces per ton; by today's standard, the ore would be worth more than $17,750 per ton.[12] The huge stopes dug out to retrieve the ore were basically underground caverns supported by timbers or by pillars of dirt not removed during excavation, even though valuable, but left to provide support to keep the ground from collapsing.

The early mining in the Quartette, and in all of the mines in Searchlight, was performed by hand. Two methods of drilling were used. The first was single jacking: one man with a large hammer simultaneously hit and turned a sharpened piece of steel. The other method was double jacking: one man held a large, long-handled hammer or mallet with both hands, striking a piece of steel that was held and turned by another man. After the holes were drilled, dynamite was packed into the cavities; a cap attached to a fuse was lit, causing the cap to explode and ignite the dynamite charge. This same method was used in shafts and for tunneling work.

Occupational safety was almost an afterthought. Miners didn't wear hard hats in Searchlight until World War II. They wore cloth hats with a mount on the front upon which to hook their carbide lanterns. Carbide is a binary compound that produces an ignitable gas when combined with water, thus allowing miners to see underground.

After Hopkins purchased the Quartette, the work gradually became mechanized. Gasoline combustion engines were used to power hoists for removing muck and ore from the shafts. Hand power was used to tram the material to the shaft from the various tunnels—drifts, crosscuts, winces, and raises. This waste and ore was placed in cars and trams that ran on iron tracks laid like a miniature railroad. At the shaft, the bucket or tram was put on skids and hoisted to the top.

In the smaller mines, the ore and waste products were brought to the surface by various means, the cheapest being a windlass. A windlass was normally a rounded wooden shaft with a crank on one side end, which had the rope or cable wound around it. When the crank was turned, the rope or cable wound around the shaft, bringing the materials to the surface. Other more elaborate hoisting methods used horses or mules to turn the crank and bring the earth up to the surface.

Even the quarters for the mine bosses at the Quartette were impressive. In 1905 new quarters constructed for the superintendent and other supervisory personnel included lounging and reading rooms. Colonel Hopkins had a complete private suite, even though he spent most of his time outside the district, in either Los Angeles or Boston.

It was reported in 1905 that even more modern provisions would come to the depths of the mine, in the form of new drilling equipment. A new compressor on the surface would supply a new drilling apparatus for drilling uppers, making it easier to place drill holes on the upper reaches of the adit. This method replaced the single and double jacking for much of the work in the Quartette. About the same time, a small timber mill was installed, including a tip saw, swing, cut-off, and wedge saw for the preparation of the mine timbers.[13]

When the main shaft reached the 800-foot level, the modern hoisting equipment allowed the skip, which held three thousand pounds of ore or waste, to go to the bottom and back to the top in three minutes. The hoist was operated by a 60-horsepower Fairbanks-Morse engine, at the time the largest made in the world.[14] Despite all the expenditures for supplies and equipment, it was determined in the summer of 1905 that it cost only $5 per ton to mine and process the Quartette's ore.[15]

Throughout its entire period of operation, the Quartette required timbers in large quantities for square-set timbering. The square-set process was invented by Philip Deidesheimer, who was brought to Virginia City during the Comstock era to solve the extremely dangerous problem of cave-ins, which frequently caused injury and death. He developed the system in just two weeks. His plan was to frame timbers together in rectangular sets, each set being composed of a square base, placed horizontally, formed of four timbers, sills, and crosspieces from four to six feet long, surmounted at the corners by four posts from six to seven feet high, and capped by a framework similar to the base. The cap pieces forming the tip of any set simultaneously functioned as the sills or base of the next set above. These sets could readily be extended to any required height and could be spread over any given area, forming a series of horizontal floors, built up from the bottom sets like the successive stories of a house. The spaces between the timbers were filled with waste rock, forming a solid cube, whenever the maximum degree of firmness was desired.[16]

Not only did this method of timbering provide strength, but it also allowed the timbers to move with shifts in the ground. The slight shifting of the ground would twist normal braces of timbers loose, but with Deidesheimer's square-set method, the bracing remained firm. In Searchlight much of the ground required the square-set method, and experienced timberers were always at a premium.

The Quartette constantly had trouble finding a sufficient supply of wood for its timbering. In September 1905 it was reported that it became so difficult to get the timber from Southern California suppliers that the company ordered 500,000 feet of the product from the Northwest. It was a time when huge amounts of timbers were needed because the shaft was reaching the 1,000-foot level.

In addition to the timbering method of shoring up the loose and dangerous ground, many of the stopes were buttressed by leaving pillars of ore to hold the ground from caving in. In the later leasing years, even though the procedure was dangerous, the pillars of ore would be taken, leaving the ground without support.

The Quartette used timbering only as a last resort. This is clear from early statements made by Colonel Hopkins, who, when asked in 1906 if the company could take more ore than it was currently processing, replied, "It is not because we have not the desire to take out as much metal as possible in a given time, but simply because we are compelled to protect our mine from the possibility of collapse owing to the character of the walls. With an increased output it would be for us necessary to do much costly timbering to keep the mine from caving in that it would not be worthwhile, whereas at present we are safe from disaster and are doing very well indeed with our investment. In the course of time we will reach a stage where we can work upward and then will be asked to mine on a larger scale."[17]

By June 1905 the Quartette had already produced more than $800,000.[18] Before the end of the same year, the mine would have produced more than $1 million,[19] a huge sum of money for just after the turn of the century. In August 1905, 325 men were employed in the mines in Searchlight; this figure did not include the many supporting workers such as teamsters, millers, and the businesses that supported the town and the mine workers. Seventy-five of these men were employed in assessment work and by contract—that is, they were not employed for wages as other miners were.[20] By far the largest employer in the county was the Quartette Company.

As late as 1908 there were those who wrote that because gold was still present at depths of nearly 1,000 feet, the mine would have a virtually inexhaustible supply of good ore.[21]

About the same time that the Quartette's river mill began operating with ore supplied by the company's own railroad, water was hit at the mine. In fact, one of the interesting phenomena in the Searchlight area was that some of the mines hit water at relatively shallow depths. The Santa Fe, located about a mile and a half from the Quartette, found water at less than a hundred feet. The Quartette didn't hit water until about the 500-foot level. The local newspaper reported: "It is supposed to be the scarcest article in the desert, but mine after mine here is developing water in unheard of quantities."[22] Even though the water came at relatively deeper levels in Searchlight, when water was reached, it appeared in large quantities. At the beginning of 1908, the Quartette was pumping 200,000 gallons a day out of the mine.[23]

The dewatering of the mine allowed the company to build a mill closer to the mine site. By October 1906 the twenty-stamp mill was crushing 2,000 tons monthly. Like the rest of its operation, the mills of the Quartette were state-of-the-art facilities. By the end of the year the company had added another full twenty-stamp mill and was then milling more than 4,000 tons each month.[24] These mills were used well into the 1920s before they were replaced by ball mills, which were much more efficient and less costly, requiring significantly less maintenance.

By the summer of 1909 the 1,200-foot level had been reached in the main shaft. In August ore of a very high value was found in one of the drifts at the 1,100-foot level. At the same time a new ore body was announced at sites between 400 feet and 500 feet down in the workings.[25]

The bowels of this magnificent mine were extraordinary. Even as early as 1906, the description of the mine was inspirational: "It would take several hours to make even a hurried trip through the several miles of underground workings. The mystical maze of drifts, stopes, upraises, crosscuts and winces confuses one . . . and the visitor simply loses what mental balance he has left and becomes simply a human exclamation point and ejaculates an endless strings of Oh's and Ah's."[26]

From the main shaft extended various drifts, nearly horizontal mine passageways driven on or parallel to the course of the vein. On the 200-foot level the drift west was driven more than 1,000 feet. This same distance was cut on ten other levels, the 300- through the 1,000-foot levels, as well as the 1,200- and 1,300-foot levels. Cut from these miles of drifts were huge upraises of stopes, some more than 150 feet high and 150 feet wide. The most extensive stoping took place on the 200-, 300-, 1,000-, and 1,100-foot levels off the main shaft. Naturally, the deeper the mine went, the more hazardous the ground became.[27]

It is difficult to imagine the danger and hardship of working in these huge caverns. The only preserved account of the adversity came in 1934, from someone who had been in the Quartette in 1912: "The temperature was at 105 degrees, at the 1,200 and 1,300 foot levels, with the ground being very soft. The working conditions on the east face of the 1,200 and 1,300 foot levels were almost impossible even though the ore was still good. The work at almost all levels was most difficult because the stoping had been done improperly."[28]

This letter was written many years later, when Charles Jonas, formerly the superintendent for Hopkins and a subsequent lessee, was attempting to get financing for the mine. He had firsthand knowledge of the operation because he had been involved in the mine since at least 1912. Jonas observed that ore was removed in such a manner that no others would later be able to work the mine in the area where the stoping had occurred. Not only was the ground bad and the underground workings hot, but miners were also constantly fighting the never-ending encroachment of water. As late as the 1940s, residents of Searchlight could still feel and hear the Quartette's big stopes caving.

In 1909 the great Quartette Mine was still producing $500,000 a year,[29] but even as early as 1908, there were rumors that the mine was beginning to fail, and the owners were reported to be negotiating a sale to an English syndicate for $4 million.[30]

The demise of the Quartette began when Colonel Hopkins decided he wanted to turn the management over to others. In January 1910

Hopkins's son, Walter, became the assistant mine manager. Immediately afterward came the first mention of leasing out operations, even though the reports showed that the mine was doing well. But in June, thirty-five of the forty stamps in the mill were silenced.[31]

By the end of 1911 the Quartette was being leased to many different individuals, much like sharecropping in the South. Different areas of the old mine would be mined by lessees, and the Quartette Company would receive a royalty or percentage of the ore taken out by the lessees.

From a review of the mining statistics for the year ending December 1909, the figure for Clark County, almost 12,000 tons, basically referred to mining in Searchlight—no significant mining activity had gone on elsewhere in the county during the preceding decade. For the same period in 1910, the tonnage dropped to 2,400 tons. The main obstacle to further success was the extremely high cost of taking ore from such deep areas of the mine. It is clear that the leasing emerged for primarily economic reasons.

By July nine lessees were operating above the 100-foot level in the Quartette. It was said that the mine was a leaser's mecca because the lessees had some good luck reworking the tailings.[32] Most of the work was at the upper levels of the mine, with some miners sinking new shallow shafts. By the end of 1922 a significant amount of work was being conducted near the surface of the old glory hole, the shaft Macready had opened to start the Quartette.[33] In January and February lessees hit ore at 20 feet, 40 feet, and 1,350 feet.[34]

Most of the mining camps in Nevada experienced much the same evolution as Searchlight, with leasing following the initial production. Tonopah, however, was unusual in that the leasing came first. Within a year of the initial discovery of gold in Tonopah, Jim Butler, the discoverer, had granted more than a hundred leases on his property. He received a 25 percent royalty on the production of the ore. In Searchlight the formation of the large mining companies came shortly after the discovery of the valuable minerals. In Tonopah the large companies came after the leasing era.[35]

Some believed that the labor unrest of 1903 encouraged miners to secrete certain valuable ore deposits during and shortly after the strike.

This information was a good basis for the mystic mind of the miner who envisioned hidden treasures of gold deposits. In the report to investors in 1934, Jonas would write, "The prior leasing operations success depended upon the secret knowledge held by certain people who had secured this information from the unscrupulous group who operated the mine prior to 1905."[36]

Leasing of any consequence at this great mine was basically concluded by 1917. The success of the lessees is not fully known. Though some miners did quite well, most made insignificant profits. Montgomery-Jones earned $40,000; Post, $20,000; Holmes-Jones, $80,000; Hockbee, $15,000; Pemberton, $5,000; Hudgens, $40,000; and McCormick, $40,000. The discovery of new ore deposits was negligible, with most of the value coming from the recovery of ore left by the Quartette Company for safety reasons. The lessees would simply remove the dirt pillars, causing further degradation of the mine. Several had good luck near the surface, such as John Hudgens, who worked the surface east of the air shaft. He removed $40,000 of ore from the Quartette by going after some ore left behind by the Hopkins group. But years later, in 1931, his son and grandson obtained another lease on the Quartette. They took out about 60 tons that assayed at $50 per ton. This was a good find considering it came from an area no larger than twenty feet by fifteen feet.[37] McCormick removed his value on the 600-foot level at a point of a drift 400 feet from the main shaft. This block of ore was deliberately hidden by the crooked management of the 1903 era and ran more than $200 per ton.[38]

The mine that made Searchlight would continue to be excavated for many years to come, but the glory hole was rendered unusable, as were most all of the areas in the Quartette that had been worked before 1917. The mine, with its large caverns, was too dangerous even for the most courageous and, at times, foolhardy miners. Most of the work in the future would be promotional at best; never again would the magnificent mine produce ore of any consequence. But neither did any of the other mines in the district.

7

THE COMING OF THE RAILROAD

Three major events contributed to the optimism among Searchlight's founders in the early days of the new town. The first was the almost miraculous discovery of the big vein in the Quartette. The next event that had long-term consequences was the union work stoppage and subsequent lockout by the owners at final settlement. The final significant event was the coming of the railroad.

Like most western pioneer towns at the turn of the century, Searchlight employed the typical modes of transportation for freight and passengers. First, pack animals, which moved along trails, were used. Then came the more modern conveyance, the stagecoach, powered by both animals and machines, which hauled people and freight and traveled on better roads, the forerunners to the highways now used by contemporary automobiles. Searchlight initially benefited from both of these types of transportation. The railroad would come later.

Trails and roads quickly linked the new town to the Colorado River settlements and other nearby towns, such as Nelson, Needles, Barnwell, and Nippeno (which became Nipton). As mentioned earlier, the Eldorado Canyon Road had been blazed at least thirty-five years earlier, passing near the Searchlight site. The first stage line that served Searchlight on a regular basis was driven by the owner, H. C. Bartees, in 1898. At the time he delivered his first load into Searchlight, he found three buildings in town: Roses' Cottages, Black's Store and Boarding House and Lunch, and finally, Crawford's Saloon. Black's Store was the first building in Searchlight and was opened in July 1898.[1] In December of the same year the first post office was started, and S. H. Black was named the town's first postmaster.

Like many western mining camps, Searchlight was located far from other places of commerce. It was not, however, as remote as some of the other Nevada gold and silver boomtowns. Goldfield, Tonopah, and Ely, for instance, were even more isolated than Searchlight.

The unusual, sometimes humorous, and sometimes changeable names of the communities around Searchlight area often created confusion. For example, twenty miles west of Searchlight, near the end of the railroad track, was a flourishing town that the Isaac Blake Company named Manvel, for Allen Manvel, who was president of the Santa Fe Railroad. The Santa Fe line reciprocated by changing the name of Goffs to Blake in March 1893; then in November 1901 it was changed back to Goffs. The name Manvel was changed by the Santa Fe to Barnwell in 1902. To further complicate matters, the railroad and the post office didn't change the names of the two places at the same time. No one seems to know how Goffs got its name.[2]

Searchlight may be one of the few towns in the world during this era where the automobile preceded the railroad as a mode of transportation. The first automobile service was established in late 1903. J. R. Bob Perew organized an auto stage into Searchlight. He came from Barnwell, initially in a 16-horsepower car pulling a two-wheeled trailer. From the first very long and difficult trip in September he was able to cut the time of the journey to less than two hours.[3]

Another stage line also saw service before the coming of the railroad. The Searchlight Stage and Freight Line started service in 1904, establishing the first regular travel to Nipton from Searchlight in 1905. This company initially competed with the Perew operation for the Barnwell route. Its first trip from Barnwell was a failure because the vehicle high-centered on the road and rocks damaged the undercarriage. The Barnwell route was used more frequently than the Nipton road because the road was already well established, as was the trade on the road. The Nipton road was not regularly traveled until the San Pedro, Los Angeles, and Salt Lake Railroad line was opened in 1904. By March of that year the newspapers carried advertisements from the Searchlight Stage and Freight Line for regular trips, to Barnwell and also to Eldorado Canyon, by truck and car.

With the entrance of the Salt Lake line in 1904, a new route was available that not only opened a new travel lane between Los Angeles and Salt Lake City but also made accessible new service to Searchlight through Nipton and Leastalk (a name developed from rearranging the letters in Salt Lake; the town later became Ivanpah, an Indian name meaning clear or good water).

The *Searchlight,* in the spring of 1905, wrote that the town of Searchlight would have at least 2,000 residents soon, because just a year earlier the town had had only one railroad with service on a tri-weekly basis, and now the service from both the Salt Lake and the Santa Fe lines would be passing by on a daily basis. The newspaper was misleading its readers—the town didn't actually have its own service but relied on the distant Barnwell Railroad, which served Searchlight by stage.

The nearest railroad connection in the first few years of the town's existence was the station at Barnwell, the northern extension of the Atchison, Topeka, and Santa Fe. This line provided the best link to the outside world, as there was a train from the main line to Barnwell three times each week. Passengers could then transfer to other modes of travel to come the rest of the way to Searchlight. The ride from Barnwell, a distance of 23 miles, or from Nipton to Searchlight, a distance of 22 miles, cost as much as the journey from Los Angeles to Barnwell or Nipton, a trip of more than 200 miles. To travel to Searchlight from Los Angeles, a passenger would leave the California city in the evening, arrive at Barnwell at noon the next day, then depart by stage for Searchlight at 1:00 P.M., not arriving until 6:30 in the evening, almost twenty-four hours after leaving Los Angeles.

The first operating railroad in Searchlight was the fourteen-mile narrow gauge from the Quartette Mine to the river. The experienced railroad builder, Colonel H. W. Dunn, was the builder as well as one of the original developers of the Quartette. When word of the railroad filtered up to Lincoln County, the editor of the Pioche newspaper sarcastically remarked: "I suppose in the near future, a line of ocean steamers will be run in connection with the railroad for the purpose of conveying gold bars from the river mill to Boston and then beans back to Searchlight."[4] This statement reflected not only the disbelief of the northern part of the county but also its jealousy of the economically stimulated southern part of the county.

The construction of this railroad was not exempt from problems or complications, however. The eighteen-ton locomotive and other equipment for the new railroad and mill were shipped up the Colorado River on a barge, which got stuck on one of the river's sandbars, below the site for the mill, and remained there for three months, visible from the mill site. The heavily loaded barge was not freed until the Colorado flooded in February. After this experience, the construction crew gave up on the river and had the rest of the rails and other equipment brought in from Manvel by wagon train. Shortly after the barge was freed from the sandbar, the locomotive was available to assist in the construction of the rail bed and haul supplies and workers. The locomotive burned both oil and wood; the oil was freighted into Searchlight, and there was ample driftwood on the banks of the river.

The grading of the line began in November 1901, and the track was opened in May of the following year.[5] In comparison with the earlier setbacks, the grading proved to be a relatively easy part of the construction project. After the track was completed to the Quartette, the train

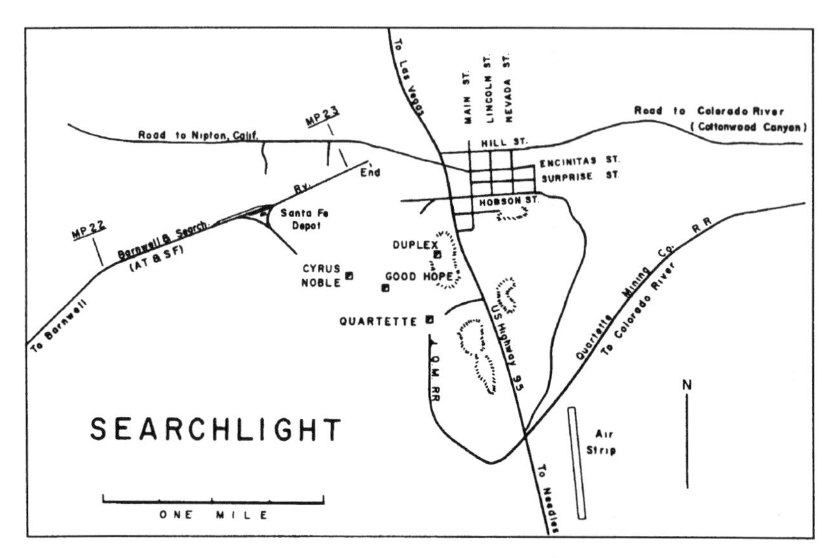

Searchlight and its railroads. Map by Robert B. Adams. From *Railroads of Nevada and Eastern California* by David F. Myrick, copyright © 1963, 1991, 1992 by David F. Myrick. Used with the permission of the author.

occasionally brought supplies to the town from the river. This freight came from the nearby farms on the river and also from the river steamers. Several young engineers also rigged a handcar with sails and established the Desert Yacht Club to cruise up and down the line.[6] This small railroad also handled passengers who wanted to catch one of the steamers that used the Colorado to go downriver and thus contributed to the boom that occurred during the first few years of the first decade of the twentieth century.

The river supplied the one element that was uniformly scarce in the desert—water! Not only was there water, but the entire fourteen-mile stretch of track down to the river had no hills. Therefore, trains loaded with ore were always heading downhill, and on the return trip uphill the freight cars were lighter and generally carried water and produce.

Water shipments from the river were essential to the water-intensive milling process at the mining camp. As the standard-bearer for gold and silver mining production, the Quartette Company built a twenty-stamp mill at the river site. In less than two years, however, after water was discovered at the mine site, the narrow-gauge railroad was no longer needed, and by the end of 1903 it had ceased to run.

The tracks of the narrow gauge were sold in parts to various enterprises. In March 1907 the Yellow Pine Mining Company in Goodsprings bought some of the steel.[7] Other parts of the narrow gauge were purchased by a Las Vegas gypsum mining company. The mill was disassembled and brought to the Quartette, which doubled its number of stamps from twenty to forty.

After the construction of this first rail project, increasing attention was given to the development of a full-gauge railroad, with a terminus in Searchlight, that would connect to one of the major rail lines. The first promoter of such a line was Colonel Hopkins of the Quartette, who was not only persistent but remarkably proficient.

Initially, the first public conversation concerned the Ibex station on the Atchison, Topeka, and Santa Fe line, Ibex being located about forty miles south of Searchlight. This route was discussed as early as 1901 and shortly thereafter was surveyed. Later, in 1904, a rumor circulated in town that a freight house and other buildings were being constructed at Ibex and that the Santa Fe was considering a direct line into Searchlight. In 1905 an official from the Santa Fe line came to the district, prompting speculation that there was going to be a new line called the Needles, Searchlight, and Northern, with Searchlight as the terminal, which would then be connected with Eldorado, Crescent, and other mining regions.[8] There was no validity to these rumors.

The most important factor in the eventual construction of a railroad into Searchlight was the establishment in 1904 of the new route between Los Angeles and Salt Lake by the San Pedro, Los Angeles, and Salt Lake Railroad Company, creating possibilities for new service to Searchlight through Nipton and Ivanpah. The Santa Fe group immediately feared that a competing route could be built relatively easily between Searchlight and Nipton, since the nearest point in the Searchlight mining district was only sixteen miles away.

A review of the internal company correspondence allows a glimpse into the events that stimulated the construction of the line between Barnwell and Searchlight.[9] The first telegram from E. P. Ripley, an official with the Santa Fe, dated March 9, 1905, alerted his boss, president of the line Victor Morawetz, as to the location of Searchlight and his fear that the Salt Lake line would divert the Searchlight mine business, which at the time was the exclusive province of the Santa Fe. Ripley also indicated his preference for the Ibex site.

On March 11 of the same year, Hopkins wrote to Ripley requesting that the Santa Fe underwrite bonds or in some other manner assist the Quartette group in constructing the line. Hopkins wanted to use the equipment from the narrow-gauge line for the new line connecting Searchlight to the main rail system. This information was sent immediately to Morawetz, with Ripley's sardonic observation: "Any man who in these days builds a narrow gauge railroad over the desert where construction is easy as it is in this case is a subject for a lunatic asylum." The suggestion of a narrow gauge was immediately rejected.

Surprisingly, Morawetz suggested on the same day, March 17, that the company should consider a route to Searchlight from Chloride, Arizona. This route is mentioned several times in the company records, but never was there discussion of the obvious need for spanning the Colorado River, which suggested that Morawetz did not fully understand the ramifications of the proposed course.

On March 22, 1905, Morawetz for the first time directed Ripley to assess the extent of the ore in the Searchlight area. This seems an obvious issue that surely should have been dealt with earlier. On the same day Ripley responded that "examinations that had previously been conducted had been promising but not of a character which would warrant us building a road." This is consistent with the first telegram, of March 9, when Ripley correctly informed Morawetz that the ore in the district was of a low grade. Searchlight did have a few quality ore deposits, but they usually appeared in very small outcroppings.

Always underlying the Santa Fe's communications was the fear that if the company did not proceed in some manner, the Salt Lake line would take advantage of its inaction. This concern was mentioned many times in the company documents, as when Ripley suggested that "if they fail to deal with us then the connection will be made with the Salt Lake road."[10]

One of the fears raised by Victor Morawetz was that the push to build the railroad was only a subterfuge for the speculative promotion of the town and was not based on an urgent need to provide expanded transportation facilities for the low-grade ore. On March 28, Morawetz again queried Ripley: "If we do nothing, a line will be built out from the Searchlight district to the Salt Lake Line. Everything seems to me to depend upon the question whether the Searchlight district is a promising mining camp and whether the Searchlight people believe in their property and want to build the road for the purpose of working it and not for the purpose of selling it. Can you not find out what they really have in view? It would be a rather clever scheme on the part of these people to get a road built out with the assistance of the Atchison at a minimum of cost to themselves so that they could use our action as an endorsement of their enterprise with a view to unloading it on others."

Ripley could not make up his mind about assessing the strength of the mines in Searchlight. Morawetz had accessed information that George Gould, the famous industrialist, had investigated Searchlight and found it inadequate for future investment. Ripley indicated that he was not familiar with the report but was not surprised that such a report existed. He explained: "Our experts decried the value of this district when they first examined it about three years ago, but on a later examination within the last six months the same examiner admitted that much more had been accomplished by the mine owners than he had before considered probable and while not fully endorsing the district, said that the Searchlight mine with proper transportation facilities would probably produce a large low grade ore tonnage, though he still believed the high grade ore was limited in quantity."[11]

Within the company there was much vacillation. The decision to build a rail line into Searchlight was not an easy one. Though the ore in the district was generally of a low grade, the mines in the area were producing lots of it. The decision was also being driven by the competition from the Salt Lake line, as well as various promotional schemes to finance the construction privately, to either the Santa Fe or the Salt Lake system.

The situation was summarized best in a telegram from Morawetz to Ripley, on April 6, 1905: "The problem is a difficult one. It is one of those problems where you cannot be certain but must take some chances, and I am quite content to leave the matter to your best judgement, as you are on the spot and can take every element of the problem into consideration."

Given this new authority, Ripley began to consider several different and competing scenarios. He reviewed different bonding proposals, several partnership options, and a number of private proposals. But almost a year passed before the company took any definitive action.

Both Searchlight's newspaper and Colonel Hopkins believed the area's development would be enhanced if both the Santa Fe and the Salt Lake lines were built. Everyone in the railroad business knew, however, that the district would be lucky to be able to support one line, never mind two. At one point, in a effort to stall others who appeared to be moving on the Salt Lake connection, the Santa Fe management ordered a survey crew to begin work. The bluff worked; the newly formed Needles, Searchlight, and Northern did not proceed with its announced grading, which was to have begun on March 23, 1906.[12]

After years of indecision, in early April 1906 railroad officials announced that the Atchison, Topeka, and Santa Fe would begin construction of the Barnwell route, with completion anticipated by August or September of the same year. Construction actually began in May.

The railroad moguls formed a puppet corporation with the name Barnwell and Searchlight Railway Company, which was officially incorporated on April 16, 1906, in the state of California. Immediately thereafter, this company was leased to the Atchison, Topeka, and Santa Fe.

The Searchlight newspaper covered the construction on a weekly basis. Progress was exceedingly slow. The work fell well behind schedule because of severe labor shortages, and the workers who were on the job had insufficient experience. The terrain and weather also proved much more difficult than anyone had projected.

On March 15, 1907, the newspaper disparagingly reported that the work crew included more than 2,000 Mexicans, dressed in gaudy sashes and big sombreros. The paper's prejudices were evident in its report that these laborers worked slowly and sang songs in Spanish, which the paper thought unproductive. Few men, including the Mexican workers, remained on the job for more than a week. Most left to try their fortune in one of the local mines, while some, lured by the stories of gold, began prospecting on their own. The railroad company raised wages several times, but the hard work, the poor living conditions, and the dreams of striking it rich created instability in the work crews.

The daily crews were composed of as many as 400 men. According to one of the project timekeepers, more than 7,000 men were required to accomplish road surveying, grading, track laying, and general construction.[13] Even so, it is hard to imagine how the paper could have accurately reported the total number of workers, when in reality the project was not completed until almost five months after the date of the article. The immensity of the task is illustrated by the fact that there were as many as 900 teams of animals at work on the grading alone.

Before the construction work on the line itself was completed, controversy erupted in the new mining town over where the terminal for the new rail system would be located. The townsfolk wanted it to be in the middle of the commercial part of town, but the railroad officials knew that the purchase of land in the middle of town and the extension of the grade into the hilly center of town would only increase costs. The company decided to place the terminal about one mile to the west of the main business district. The railroad announced the decision on November 2, 1906, affirming that the company had secured surface rights for a separate townsite that would have 1,250 lots for sale. In addition to the lots, the new site would have a business district surrounding the depot, a plaza, a park and bandstand, and a hotel costing as much as $7,000.[14] The Santa Fe also determined that the new location would have a new name, to distinguish it from Searchlight. The new town would be named Abbotsville, after the townsite manager.[15]

The business community in Searchlight rose up in anger against the railroad. In fact, advocates of both the Searchlight and the Abbotsville locations ran competing advertisements in the newspaper beginning on

November 30, touting the benefits of the respective sites and urging people to buy lots from them. The Searchlighters vehemently argued that, just as there was only one Los Angeles, there was also only one Searchlight. Obviously, the railroad company seemed to agree; it was, after all, proposing a different name for the new town.

Those familiar with the original Searchlight townsite and the railroad's alternate terminal location would have to side with the railroad, for the site it chose was flat and almost entirely outside of the floodplain. Searchlight was built below the Duplex Mine, and the area had many hills and small mountains, which would certainly interfere with the sound operation of a rail terminal. In spite of the railroad's reasonable position, those with commercial interests in Searchlight were very bitter because over only six years they had invested more than $750,000 in building the present business district.

In spite of the disagreement over the terminal site, the rail line was completed; the rail station included ticket and freight facilities, as well as living quarters for the agent and his family in the upper story of the depot.[16] The new depot was a showcase by any standard. Beveled and ornamental glass was used in both the first and the second stories of the building, but particularly in the upper portions of the windows, where intricate, fancy patterns were etched into the glass. The hardware on the door and windows was made of brass.

The first train, as a test run, arrived on March 31, 1907, seven months behind schedule. The next day the first regularly scheduled train arrived, carrying eighteen passengers. The real festivities occurred on April 13, when railroad day was celebrated and nine coaches arrived, pushed by three engines and carrying more than three hundred visitors.[17]

This two-day celebration was proclaimed with a large banner that read, SEARCHLIGHT: THE CAMP WITHOUT A FAILURE. The celebration included a parade, a barbecue, a drilling contest, a band concert, and even tours of the mines and mills. Those who came on the excursion train represented various special-interest groups, including the Los Angeles Merchants Association, the Los Angeles Chamber of Commerce, and the Los Angeles Stock Exchange.[18]

The *Bulletin* often wrote in exaggerated fashion about the economic potential of Searchlight, but it stretched the limits of hyperbole in referring to Searchlight as an important part of a transcontinental railroad.[19]

The train service brought many predictions of exaggerated urban growth in the train service area, at or near Searchlight, but in fact the new railroad lost money every year of its operation, although the first three years did provide a steady supply of freight.[20] The completion of the line initially brought a boost to Searchlight, but unfortunately about that time, the Quartette began to cut production because of changes in management structure and the lack of good ore. The town continued to boom for a few months, but it was evident that mining operations were declining. Even after the mines stopped shipping large quantities of ore, however, the trains were still used to ship cattle out of Searchlight, and refrigerated cars brought in supplies to the town's stores and markets.

The Santa Fe Company continued operating its Barnwell-to-Searchlight run, on Mondays and Fridays, until the Interstate Commerce Commission granted permission in 1923 to abandon the line, following a severe washout.

The railroad had a tremendous impact on Searchlight even after its demise. The railroad ties remained in place, and Searchlight residents for years to come extracted the ties for firewood. They burned well, since the wood had been soaked in an oil substance that prevented rotting. People also used the ties in building homes—my own family lived in a home built entirely of the railroad timbers, which, though not fire-retardant, resulted in a very stout house. Ranchers also used the ties for fencing and corrals.

It was an ignominious end for a project that fewer than twenty years earlier had held such promise and raised so many expectations.

8

A FASHIONABLE TOWN

One of the best descriptions of the businesses in early Searchlight comes from a petition signed by the operators who opposed the Santa Fe Railroad's locating the new townsite outside of the boundaries of the original townsite. Contemporary newspaper advertisements also provide a portrait of the business activities in the town.

A news article in the local paper in the winter of 1907, when Searchlight was in a period of rapid economic growth, boasted of the advancement of the "town without a failure," noting that Searchlight had "a city water works, ice manufacturing and cold storage, telephone plant, electric light plant, one bank, one newspaper and job printing, three lumber yards, seven hotels, six restaurants, nine saloons, one wholesale liquor store, two barbershops, one bath house, four general stores, one grocery, one meat market, one bakery, one gents furnishing and clothing, two drugstores, one leather and harness shop, one watchmaker, one dental office, three physicians' offices, one hospital, five lawyers, one district recorder's office, one Justice of the Peace, two deputy U.S. mineral surveyors, ten real estate and mine brokers, six mining engineers, two assay offices, one tent factory, two liveries, three laundries, three carpenters' shops, three stage and freight lines, one paint house, one blacksmith and wagon shop, one school, one church, one miners' hall, one cigar manufacturer, one tin and plumbing shop, one wholesale jobber, two wood dealers, one ranch, six mine contractors, one bowling alley, one post office, one stationery store, and 168 residents, not including those connected with the mines or under construction."[1]

It is difficult to verify the existence of each of these various businesses, but newspaper advertisements, interviews, and other documentation do yield some information. We know that the first store was opened in Searchlight by S. H. Black in July 1898 and that in December 1898 a post office was established, with Black as postmaster.[2]

With the passage of time, tales of Searchlight's commercial development became exaggerated well beyond fact. For example, a 1907 newspaper mentioned the existence of nine saloons. Yet seventy years later, a Las Vegas newspaper wrote of twenty-two saloons.[3] Three years after that, in 1980, the *Las Vegas Sun* reported that there were thirty-eight drinking establishments in the mining camp by 1904.[4] There is no substantial basis for these grossly exaggerated numbers. Even during the 1950s, when Searchlight experienced a boom related to prostitution, there were never more than thirteen bars or saloons in the town.

There is, however, in all the news of the early part of the twentieth century, confirmation of a significant number of hotels and other types of rooming houses in Searchlight. The most well publicized of these was the Hotel deRoulet, whose name was changed to the Hotel Searchlight before it opened. Built in the middle of the business district, at the northwest corner of Hobson and Main Streets with a frontage of seventy-two feet on Hobson and fifty-two feet on Main Street, the structure was two stories high, with a flat roof and an ornamental galvanized cornice. On the upper floor were nineteen rooms and two baths.[5] The name deRoulet came from the name of the intended original manager, Charles deRoulet. When he left before the hotel opened, the name was changed. The hotel's furnishings were purchased at a cost of $10,000, and the dining room was built to accommodate up to fifty patrons.[6] The top floor housed the sleeping rooms, and on the first floor were the bar, dining room, kitchen, and office space. The hotel had its own carbide power plant, which produced enough energy to light fifty lamps.[7] The general downturn in the local economy, however, forced the hotel to close in December 1909.[8]

Searchlight was commonly frequented by the rich and famous, all trying to exploit the town's economic potential. In 1903 U.S. senator Chancery DePue of New York came to town, accompanied by the president of New York's Coffin Bank and Trust.[9] The affluent visited Searchlight during this period in order to assess whether an investment in the boom was responsible or risky. Because of the legendary success of other Nevada mining ventures, most notably Viriginia City, investors were interested in getting involved early. Virginia City wealth sent William Stewart, William Sharon, John Jones, James Fair, and Francis Newlands to represent Nevada in the halls of the U.S. Congress. Mining wealth from Nevada was visible on the national scene, stimulating interest in any gold or silver strike in the Silver State.

The need for sleeping accommodations in early Searchlight was quite apparent. A revealing letter to the editor in 1905 lamented, "I am a travelling man, and I have a kick which I want to register. I have been making this town right along for the last year and a half and I have never before had any trouble getting a bed. This time I had to sleep on a shakedown on the floor of one of my customers and there were three other men there in the same fix."[10]

It became so difficult to rent a room that tent hotels began springing up. "In answer to the urgent demand for accommodations a number of tent lodging houses are going to be replaced as quickly as possible by wooden structures. It is a long time since there was a tent hotel in Searchlight. A month ago no one would have patronized such a place. Today the need is for beds and people are glad to get them under, even if it only be canvass. Wooden buildings can't be put up fast enough."[11] One of the wooden structures, the Star Lodging House, was not cheap, with rooms at fifty cents per night; it claimed to be the largest rooming house in Searchlight.[12] The Home Lodging House was cheaper, with a bed going for only thirty-five cents each night.[13]

Several comfortably furnished hotels were developed quickly. The Hotel Southern was advertised as the headquarters for Searchlighters. The Kennedy House, later to become the Hotel Nevada, was advertised as spacious and well appointed, with a kitchen and tastefully appointed bedrooms on the second floor.[14]

Early Searchlight also boasted what one would think would be rare in a mining camp—a tennis club. The *Searchlight* reported that tennis had emerged as the new popular pastime among the employees at the Quartette. A club was organized, and a fine court was laid out that would satisfy even the best players.[15] Surprisingly, several other early mining towns also had tennis courts, among them Tonopah, McGill, and Ely.[16]

In 1981, seventy-five years after the town's most productive period, one of the Las Vegas newspapers observed that Searchlight was about to get its first church. The article stated that previously Searchlight had been such a rough town, with all of the bars and prostitutes, that though several missionaries and evangelists had tried to start a church, the result was always the same: the church activities failed within two months.[17] This newspaper article had no basis in fact whatsoever.

What Searchlight never had, until the early 1980s, was a freestanding church building. Church services in early Searchlight, as in later years, were held in other already constructed buildings, such as hotels and schools. One of the early religious leaders in Searchlight was William Hauptmann, from Alma, Nebraska. As early as February 1905, he was in town drumming up support for his new Congregational church. By April it was becoming clear that his efforts would be successful. By summer the local paper had announced that church services, which included Sunday school, would be held in the schoolhouse in the mornings, with an evening service to be held in Hopkins Hall,[18] a great hall in downtown Searchlight that had been built and paid for by Colonel Hopkins and a Los Angeles associate, Frederick Rindge.[19]

By the following spring, the church was sufficiently strong to attract visiting preachers. Church members were still trying to find land for a permanent structure.[20] By this time they owned a new piano and were in the process of developing plans for a building.[21] There is no evidence that a church building was ever actually constructed. There is, however, ample evidence that various denominations held regular services for many years. Evangelistic services were frequently advertised, some of them held in the Kennedy House and others in the school.[22]

In the "Personal" section of the newspaper it was not unusual to find notices stating: "Church services will be held as usual Sunday evening at Perkins Hall."[23] Perkins Hall had become the site for most of the church services, including Sunday school in the morning at 10:00 A.M., morning preaching service at 11:00 A.M., Christian Endeavor devotional at 5:30 P.M., and evening services at 7:30 P.M. In addition to these traditional services, Searchlight even had a time and place for Christian Science Readings and for a regular meeting of the Christian Endeavor Society.[24]

In an effort to calm the town during the strike, and perhaps to atone for developing its own business district and housing facilities, the Quartette, through Hopkins and Rindge, constructed Hopkins Hall in downtown Searchlight. This facility, named after Colonel Hopkins, would be used for years as a spot for charitable events and other large gatherings. Hopkins Hall was described by the local paper as planned to be "by far the finest building in Searchlight. It is two stories and built of the finest redwood. On the first floor will be the library and reading room, and a billiard parlor. The assembly and dance hall is on the second floor. The hall on the second floor measures 48 feet by 26 feet, and at one end has a raised stage. Around the room stationary seats have been built. The room is finished with a dark green lining and a matched wood wainscotting giving it a very attractive appearance." The official opening of the hall was said to the most brilliant society event that Searchlight had ever witnessed. On that night the hall was filled to capacity with handsomely gowned women and elegantly dressed men, all enjoying fine music. Hopkins Hall was not paid for by the Quartette, as some believed, but by Hopkins and Rindge.[25]

Another break from the tedium of everyday life was baseball, which was always a popular spectator sport in Searchlight. In its first two decades, the town took its baseball very seriously, as did most of America. The *Searchlight* provided frequent commentary on local ball games, like the following humorous account of a game in 1904: "The most sensational play of the game was when J. V. Allison of the Quartette was running home from third base. It was a race between the runner and the ball. The crowd surged forward in fever of excitement. It was even money which would reach the plate first. Then just at the critical moment, when it looked like a sure put-out, Allison grabbed out his store-teeth and slid in with the set in his hand. He was declared safe."[26]

Searchlight had its own baseball team, which played Las Vegas, Needles, and other towns in the same region. The biggest rivalry, however, was reserved for the Quartette team. The two teams played many hard-fought games, with the Quartette usually emerging victorious. Other mining companies in Nevada organized teams too. The Kennecott Mine, which started in White Pine County shortly after Searchlight was founded, also developed great baseball teams. Because the mining at Kennecott lasted seventy-five years, its baseball teams became very good. They imported workers just to play baseball. It is likely that the Quartette in Searchlight did the same.

One of the big draws of early mining camps was prizefighting, and Searchlight staged both amateur and professional boxing events. As with baseball, the best of the matches featured fighters from the Quartette pitted against Searchlight contestants who were not employed by the Quartette. There were even well-advertised fifteen-round fights, which drew good crowds.

The town also had well-announced track meets and, unbelievably, even football games. In promoting football, the newspaper reported that the town had many former college players who were eager to play the game again. No actual games were reported in the papers, though the practices were discussed.[27]

The biggest celebration every year was the Fourth of July/Independence Day commemoration. Drilling contests were held, with both single jackers and double jackers. The distinguishing characteristics between the two were the size of the hammers and the size of the steel. In single jacking, a man would stand in front of a huge boulder, hoisting a sledge with an eight-inch handle. He would have a certain length of time, usually ten minutes, to see how far he could drive a piece of sharpened steel into the rock surface. Double jacking worked on the same principle, but the sledge was about eighteen inches long and the contestant swung it with both hands while his partner held the steel, turning it as the sledge struck. This was much more dangerous than single jacking, though many fingers and hands were crushed in both drilling procedures.

The Independence Day festivities also included fireworks, baking competitions, hundred-yard-dash contests, fat-man races of only fifty yards, potato-sack races, three-legged races, obstacle races, high-jumping contests, burro races, pie-eating contests, dances, and oratorical performances.[28]

Over the years the celebrations have changed, but even as late as the 1950s, rock-drilling contests were still part of the festivities. Since the 1970s the Fourth of July in Searchlight has meant a parade and a subsequent town picnic; the games have long since disappeared.

Band concerts were held regularly, often at Perkins Hall or Hopkins Hall. Again, the musicians came principally from the Quartette, which had its own band. There is no information as to the genesis of the musical group, but it was used frequently.

The various fraternal and social clubs of Searchlight were also an integral part of the society. There were "49er" celebrations sponsored by the Elks, and the Eagles sponsored an annual ball, as did the Masons. Later, as Tonopah, Goldfield, and Ely came to prominence, they too quickly formed units of Masons, Elks, Eagles, the Knights of Pythias, and the Odd Fellows.[29] These fraternal activities were a key outlet for the mostly male-dominated mining camps, often providing the only social activity for the camp workers other than gambling and consumption of alcohol.

Theatrical productions, both amateur and professional, were held from time to time. There were concerts and piano recitals, and the Quartette hosted a costume ball. Once the paper covered a traveling animal and bird show, reporting carefully trained dogs, birds, including geese, and even a goat. In 1908 a traveling circus stopped in town, putting on three shows in one day. It featured monkeys, acrobats, strong men, and clowns—everything but elephants.[30]

Early on, moving pictures were shown at different locations, though infrequently. For the less active there were several reading clubs and even some reading rooms throughout the business community. One significant difference between old Searchlight and the present-day town is the orientation of the main street. In early Searchlight the street ran in an east and west direction, along Hobson Street. Today the business district runs north and south along Interstate 95.

Searchlight has always had a place to bury its dead. In many cases, those who died were buried at the mine site where they worked. Graves have recently been discovered at the M&M Mine, and over the years construction crews have uncovered solitary graves all over the district.

One of the earliest plots still stands in a residential area in the middle of Searchlight. This quaint burial place has a wrought-iron fence and stone grave markers standing three and a half feet high. The headstone reads: "Annie B. Wife of John Bulloch, died December 29, 1901, age 47 years. Also our son and daughter Lily aged 14, Eddie aged 9. Death beloved ones, gone but not forgotten."

The present Searchlight cemetery is truly a desert cemetery; there is not a blade of grass in the whole two-and-a-half-acre plot. Many of the graves are unmarked, the names of the deceased long since faded from the wooden grave markers. In contrast, there are some memorable gravesites with unique headstones and stone-and-rock landscaped graves. This cemetery, which was established in 1906, contains approximately 190 graves.[31]

In the main graveyard, about 160 of the 190 graves are marked with names. Approximately 30 graves are marked but without names. Jane Overy, a resident historian, discovered about 25 additional names of people who were said to have been buried in the Searchlight cemetery but whose graves cannot be located.

There are also some unmarked graves at the corner of Colorado and Encinatis Streets, on property that was part of the townsite created in the 1970s. Clark County allowed the property to be purchased, and the graves were simply graded over.

Richard Taylor, a Las Vegas businessman and Boy Scout leader, has conducted significant research on graves throughout Nevada. In 1985 he trained a Scout troop to study the Searchlight cemetery, and he published his findings in 1986 in a book titled *The Nevada Tombstone*. Though short of detail, it was well accepted in Searchlight. The Scouts also cleaned the cemetery of some of its accumulated debris.

9

NEWSPAPERS

During two separate periods, Searchlight had two newspapers serving it at the same time. In its early years, the camp was covered by the *DeLamar Lode*. Delamar was located about thirty miles west-southwest of Caliente, well over a hundred miles from Searchlight. The *Lode* focused on Searchlight only during 1903, when strikes were going on in the Searchlight mines.

The paper that had clout in Searchlight was initially named the *Searchlight*, with a stylish banner (ORGAN OF THE CAMP WITHOUT A FAILURE), published from 1903 to 1913, when it went broke. In 1906, when a few mine failures began to show up, disproving the optimism of the newspaper's slogan, the *Searchlight* changed its name to the *Bulletin*.

The newspaper's owners said the reason the name of the newspaper was changed was because the name Searchlight was too confusing, as it was at times mistaken for a hotel or a mine or some other business establishment. The new name also reflected the objective of the paper, namely, the issuance of a bulletin of the events and the happenings of Searchlight.[1] In hindsight, however, it appears that one of the main reasons for the name change was the awareness that some of the mines were going to fail; Searchlight was no longer a camp without a failure.

The ownership of the paper was ostensibly held by a man named Perkins, but the real ownership was vested in a number of the early business people in Searchlight. The corporation was formed as Searchlight Publishing, and the shareholders were E. J. Coleman, Benjamin Macready, F. P. Swindler, T. J. Henderson, John McClanahan, H. A. Perkins, C. E. Maud, T. A. Brown, W. J. Kennedy, and James Harlan. Perkins was the president and editor, and he remained in these positions for most of the life of the paper.[2] This joint ownership served the mining camp well. When the paper went out of business, one of the owners, Benjamin Macready, moved from Nevada to Southern California, where he went into the newspaper business on a full-time basis. He sold his California publication to the *Los Angeles Times*[3] and made a sizable sum of money, just as he had done with his Searchlight mining ventures.

When the paper changed its name, it still wanted to display its faith in the mining camp, and so it purchased a new press. The owners of the newspaper continually preached that the people of the town should have faith in the community, and so they decided to practice what they preached by investing in new equipment.[4]

The second time the town had two newspapers was in 1907, when the Santa Fe Railroad's terminal townsite started a paper, which it began publishing on May 3, 1907, with J. B. Flanagan as publisher. After his newspaper failed in Searchlight, Flanagan started a paper in Hart, California, just across the state line from Searchlight. Perkins called him "off again, back again, gone again Flanagan."[5] It issued its last edition on August 30 of the same year. Flanagan's newspaper, called the *Searchlight News,* had the most modern equipment and management available. The real problem was that the paper started up at the same time as the nationwide economic crisis, which, of course, affected local advertising budgets. It also came at the same time as the beginning of the production downturn in Searchlight. The paper simply could not sustain itself, since it was able to obtain little, if any, advertising.

The *Searchlight* and its successor, the *Bulletin,* were active supporters of the town and its activities. The papers were always the town's biggest and loudest cheerleaders. Their number one editorial policy was to express confidence and optimism about all events in Searchlight. They reported one big discovery after another, most of which later proved disappointing, adding little to the town in the way of long-term economic viability. It was almost unheard of to see negative news in the *Searchlight* or the *Bulletin,* especially on the subjects of mining and the general business community. The only time the newspaper had a problem with the mine owners was during the 1903 strike.

During the tenure of the *Searchlight* and the *Bulletin,* rarely was there any partisan political writing. The papers covered both parties equally and only very occasionally would they endorse political candidates.

In later years other newspapers periodically popped up, but they would last for only short periods of time. After the *Bulletin*'s termination, the only paper to have any staying power was the *Searchlight Journal,* which ran for almost two years, beginning in 1946. This weekly carried very little local news, featuring instead mostly national stories and columns. The advertising and the infrequent local stories and columns helped to provide a historical perspective on the time during which the paper was published. Even before the *Journal,* George Corn attempted to revive the newspaper business during World War I by publishing the *Searchlight Enterprise* beginning in April 1917, but it failed within a year. There are no known existing copies of the *Searchlight News* or the *Searchlight Enterprise.*

Another newspaper appeared in Searchlight in 1951 but lasted for only two or three months. Published by A. J. McLain and prominent Las Vegas printer Marc Wilkinson, the paper was called the *Searchlight News-Bulletin.* There are no known copies of this short-lived newspaper either.

At this writing, there is a monthly newspaper called the *End of the Line.* It began printing in 1996 and is owned by Carl Weikel, the son of the deceased owner of Rex Bell's Walking Box Ranch. It has a mixture of local and national news and commentary. An annual subscription is $15.

Tonopah, Goldfield, and Ely all began their development shortly after Searchlight's gold was discovered. As an indication that their population and commerce were greater than those of Searchlight, one need only review the newspaper business in those towns. By 1905 Tonopah had not only several weekly papers but also a daily, the *Tonopah Sun.* Even Goldfield had, for a short time, two dailies, the *Goldfield Daily Tribune* and the *Goldfield Chronicle.*[6]

White Pine County also developed its newspapers at the height of Searchlight's boom. McGill had its *Copper Ore* newspaper, with Ely residents served by the *White Pine News,* the *Ely Record,* and the *Ely Mining Expositor.*

The news organs in these towns were similar to the *Searchlight* and the *Bulletin*: they were their towns' biggest promoters. The papers tended to bestow a tone of respectability on the various mining ventures.[7] In hindsight, it seems evident that much of what was written by Searchlight's early newspapers about the vast gold discoveries and other bold ventures was outrageous exaggeration.

George Fredrick Colton, purported founder of Searchlight. Courtesy Stanton Colton.

The *Searchlight*, which steamed the Colorado River below Searchlight. Courtesy Dennis Casebier, Mojave Desert Archives.

Quartette Mine, 1907, with superintendent's home in the background. Courtesy Dennis Casebier, Mojave Desert Archives.

Good Hope Mine, 1907. Courtesy Dennis Casebier, Mojave Desert Archives.

Searchlight miners at the site of a windlass, ca. 1907. Courtesy Dennis Casebier, Mojave Desert Archives.

Quartette work crew, ca. 1905. Courtesy E. W. Braswell.

Downtown Searchlight on a cold winter day, 1906. Courtesy Dennis Casebier, Mojave Desert Archives.

Hobson Street, looking west toward the eventual site of the Santa Fe Railroad depot, 1906. Courtesy Dennis Casebier, Mojave Desert Archives.

Stagecoach on the road from Searchlight to Nipton, ca. 1901. Courtesy Dennis Casebier, Mojave Desert Archives.

Narrow-gauge railroad from Searchlight to the Colorado River, ca. 1903. Courtesy E. W. Braswell.

Downtown Searchlight, ca. 1906. Courtesy E. W. Braswell.

Business and residential areas of Searchlight, ca. 1905. Courtesy E. W. Braswell.

Searchlight, Nev. 190 No.

Searchlight Bank and Trust Co.

Pay to the order of $

Dollars

(left) Bank note from Searchlight Bank and Trust Co., ca. 1907. Courtesy of Donna Jo Andrus.
(below) Interior of the Searchlight Bank, ca. 1906. Courtesy Dennis Casebier, Mojave Desert Archives.

Jim Cashman's garage (Jim Cashman at far right) ca. 1912. Courtesy Tona Siefert.

Cover and first page of "Searchlight Rag" by Scott Joplin, 1907. Courtesy of Library of Congress and Doug Davies.

Searchlight Railroad and Depot, ca. 1908. Courtesy E. W. Braswell.

Fort Piute, 1966. Rock wall showing "lookout" or "gun port" is no longer standing. Courtesy Dennis Casebier, Mojave Desert Archives.

The original Searchlight school being moved to a new location to be used as a bar, 1939. Courtesy Jean McColl.

James Cashman. Courtesy of Tona Siefert.

Searchlight Fourth of July celebration, hand-drilling contest, 1931. Courtesy Terry Hudgens Sr.

(top) Memo report on gold bullion from Searchlight deposited at the U.S. Mint in San Francisco, November 30, 1938. Courtesy Bill Kelsey. *(bottom)* Assay certificate for the Quartette Mine, 1938. Courtesy Bill Kelsey.

MEMO. REPORT ON [illegible] BULLION deposited at the Mint of the United States at San Francisco

ASSAY CERTIFICATE

August 17, 1938 1938

Wilders, Douglas & Kelsey Lease

Quartette Mine

Sample No.	Owner's Mark on Sample	Gold		
1	2nd Round in raise	.87	30	45
2	1st Round in Raise	.10	3	50
3	Ore on 200 Level	1.50	52	50
4	Face 400	1.00	35	00
5	Cut on 200 Level	.06	2	10
6	Cut on 200 Level	.36	12	60
7	1st Round Winze	.05	1	75

Assayer

Residential area of Searchlight, ca. 1939. Courtesy Joyce Dickens Walker.

10

EVEN A BANK

At the zenith of the boom in Searchlight the country was in the throes of a depression. The "silent" panic began in March 1907, a period of economic decline marked by a drop in stock prices, high interest rates, bank closings, and even the failure of some stockbrokerage firms. Speculators were wiped out by the thousands. This was true also in the mining business. The silent panic was short-lived, however, lasting only a few months. The real panic began in September and seemed unrelenting. Some of the largest banks failed, leaving cities unable to sell their bonds; stockbrokerage firms failed, suicides increased, and oil companies teetered on the brink of failure. The reverberations of the economic disaster on a national level were felt in the countryside.[1]

Not being able to see the future, the people of Searchlight were delighted to learn in the summer of 1905 about the formation of a bank in their town. Called the Searchlight Bank and Trust Company, it was to open its doors on June 20.[2]

As with most openings, the financial institution missed the target date, but opened with much fanfare on August 12. The owners proudly announced that the bank would at first provide strictly a commercial and exchange business, then would subsequently add a savings department. The initial statements also indicated that no mining property or securities would be considered for collateral.[3]

By December the bank claimed to have almost 150 depositors, with deposits totaling $30,000.[4] By the spring of 1906 the deposits and the depositors had increased.[5] The bank was truly a sign of Searchlight's prosperity.

The owners publicly announced that they were going to build a two-story bank and office building at a cost of $10,000.[6] But suddenly it was affirmed that Dr. Homer Hanson, the original promoter of the institution, had sold the bank to Homer Tabar, of Los Angeles, and George McClintock, the bank cashier. Because of the national economic crisis, Nevada governor Sparks declared a legal holiday for all Nevada banks, but Searchlight Bank and Trust refused to take advantage of it. The owners declared that the bank was doing well, even though many northern Nevada mining camps were doing very poorly.[7]

One Nevada scholar argues that the Panic of 1907 and the effect it had on banking and investment in Nevada mining were the beginning of the decline of out-of-state investment. It also effected the establishment of a new economic and political order all over Nevada, in which political leaders and financial investors would come from within the state rather than from California.[8]

After the sale, the national problems soon overtook the bank in Searchlight. Currency was becoming scarce, and by mid-December 1906, bank officers were meeting with depositors to consider how to deal with the lack of currency. Bank officials told the depositors that they had $10,000 in local loans, and they proposed paying them in scrip up to 70 percent of the amount of deposit. This brought about a demand by the depositors for a committee to examine the affairs of the bank.[9]

The committee, appointed with the faithful Benjamin Macready as chairman, issued a startlingly negative report, recommending that the state bank examiner be notified of the bank's poor state of affairs. The report stimulated the immediate intervention of the state, and by the end of the year the bank was officially declared insolvent.[10] Times had changed quickly from the days of celebration when two large freight wagons had come to town with a big safe, fixtures, and other equipment and supplies for the new bank.[11]

Coincident with the difficulties of Searchlight Bank and Trust was talk of the formation of another bank. The speculation was not taken seriously because the investors did not live in the area and few had likely even been to Searchlight.[12]

The bank's problems prompted allegations of more than negligence. Criminal activity quickly became a subject of discussion. Homer Tabar, the president, and S. K. Williamson, a cashier who had followed McClintock in the position, were apprehended in California and subsequently placed under arrest. The charges of mismanagement and fraud were Nevada crimes, however, and after several years of intense legal maneuvering, the two avoided extradition and thus never stood trial. Williamson eventually became ill and turned state's evidence against Tabar. Tabar's attorneys filed a writ of habeas corpus and other delaying motions; a writ was granted and Tabar was released. The basis for the dismissal in California centered on the fact that the charges stemmed from crimes committed in Nevada, not in California.

Tabar was immediately rearrested, and the court again ruled against the State of Nevada, whereupon a new indictment was brought. By this time he was not available to be arrested again; he had gone into hiding. He disposed of his assets and was never brought to justice for his activities in the Searchlight banking business.

The legal question continually raised in this case was whether one could be held for defalcation in connection with directing a bank if the accused had never performed his duties directly from Nevada but had been operating in California.[13] Because of these legal technicalities, Tabar never came before the Nevada bar of justice.

Macready did his usual fine job and was ultimately appointed the receiver for the insolvent bank.[14] More than a year after his appointment, Macready came up with a number of assets for the depositors. He felt confident he could get the depositors at least a return of ten cents on each dollar invested. He was praised: "Much credit is due Mr. Macready." When he was appointed, the affairs of the defunct institution were so hopelessly tangled that there appeared to be practically no assets. But Macready tackled the job with his characteristic energy and thoroughness. The fact that the court met only twice a year delayed matters, but step by step he ferreted out the crooked ways of Homer Tabar and gathered up the scattered ends one by one.[15]

The final dividend came in May when the court, under Macready's direction, ordered a return of not ten cents but twenty-five and a half cents on each of the depositors' accounts. So not only was Macready responsible for the Quartette, the newspaper, the phone company, the development of the Santa Fe, and the town's water supply, he also recovered a significant portion of the innocent depositors' money from the only bank ever to exist in Searchlight.

11

BOOMED OUT

Starting in 1837, the price of gold in the United States was $26.67 an ounce. During the boom years in Searchlight, that value remained stable. The price was not raised until 1934, when it was set at $35 per ounce. There was no fluctuation in the price, such as we see in today's volatile floating market. So since the gold ore being mined and discovered in Searchlight and surrounding areas was of low value, it was very difficult to make a profit. Because of this situation and because the Quartette's high-grade ore was scarce, the son of Colonel Hopkins stuck with the Quartette only a few years after his father's death.

In order to disengage from the responsibility of running this massive mine so that he could return to the peace and splendor of his Eastern roots, Hopkins's son leased the property to one of the Quartette's mining engineers, Charles Jonas, a Yale-educated mining engineer who was certainly familiar with this great mine.[1] Jonas took advantage of his inside knowledge and quickly moved to take some of the remaining good values. He controlled the master lease but subleased to many others. Jonas mined the best of it for himself and took a royalty from the sublessors on other areas of the Quartette.

After two years, the younger Hopkins tired even of collecting the royalties from Jonas and sold the entire Quartette property to him. Jonas became acquainted with one of the leading businessmen in Searchlight, a Harvard graduate named B. F. Miller who operated the Searchlight Mercantile. Jonas appreciated Miller's business acumen, but he was even more attracted to his sister, who, incidentally, was a graduate of Stanford University. Jonas later married Miller's sister.

These two Ivy leaguers decided to form an equal partnership, combining Jonas's mining properties and Miller's mercantile. Later Jonas sold his interest in both to Miller and left Searchlight to become the chief operating officer of S&S Shock Absorber Company, a large company doing business in twelve western states, with headquarters in Los Angeles.[2]

Miller was personally involved in the mining and the mercantile businesses in Searchlight until 1915, when he married a Salt Lake woman, moved to Los Angeles, and entered the banking business. Heirs of Miller still own the Quartette.

After 1908 the town went downhill quickly. No matter how the *Bulletin* tried to promote the town whenever a new mining venture appeared on the horizon, it couldn't reverse the depressed economic trend. The town with its own water system, telephone company, electric plant, and ice plant was no longer expanding but was simply trying to hang on to what it had.

In most cases of mining camp development, the railroad would start cutting service when a recession or depression hit. Just the opposite occurred in Searchlight. The railroad was completed just at the beginning of the mines' lower production. Ordinarily, this would have greatly improved the profit margins of the mines. But in Searchlight's case, the new railroad could not compensate for the lower demand for gold, the Panic of 1907, and the poor quality of ore.

Bob Kerwin, a man who spent seven of his boyhood years in Searchlight, described how quickly the most modern town in Nevada faded. By the time he left Searchlight in 1915, the mines were quiet and virtually all of the old productive ones, including the Quartette, Blossom, Good Hope, and M&M, had been leased, while the Duplex was still being operated by Colton. Kerwin saw the only drugstore go out of business and the pharmacy trade transferred to the Austin Store, which also had a soda fountain. There were two other stores, Searchlight Mercantile and the Pendergast Store, which were both very small general merchandise operations. Left over from the boom days were five rooming houses. They were all fairly small except Wheatly House, which served three meals a day.

It is from Kerwin that we first hear of James Cashman, one of Searchlight's most noted residents. In 1915 Cashman had a garage and service station. Even though he left Searchlight in 1920, the signage on his tin garage was legible well into the 1960s. Cashman would later own the Searchlight phone company. Kerwin recognized that Cashman, even as a young man, had an unquenchable thirst for the up-and-coming age of the automobile. Cashman showed the residents of Searchlight that his Hudson would go forty-five miles an hour, and he did this on more than one occasion. His exhibitions took place on the dry lake just south of the present-day railroad pass.[3]

Jim Cashman was a dealer in Essex and Hudson cars, operating the first automobile dealership in Nevada.[4] He also proudly advertised his Searchlight Ferry, which connected Arizona with Nevada on the Colorado River just below Searchlight. In his advertisements, he also promoted his stage line, with service to Nelson and Las Vegas twice each week. This advertisement ran in the *Las Vegas Age* newspaper, in 1921, the year after Cashman left Searchlight to take up residence in Las Vegas.

The youngest of eleven children, James Cashman came west from his birthplace in Missouri, following his older brother Harvey. He came first to Las Vegas, where he set up a tent hash house. That business did not prosper, so he again followed Harvey—to California and eventually to Searchlight.

Cashman started managing the phone company when he first came to town. Subscriptions were falling off, and after four months he began leasing it; two years later, about 1910, he purchased the company himself.[5] Cashman proved to all early on in Searchlight that he had not only good commercial power but also physical power. In 1912 the *Bulletin* reported: "James Cashman had a narrow escape from being seriously injured while working alone at the Quartette cyanide plant. In some manner his arm caught in the large conveyor belt that carries the tailings, and he was drawn into a pulley. However, his strength and the bulk of his body proved sufficient to partially stall the belt, which is 21 inches wide, thereby releasing himself."[6]

Cashman left Searchlight to become a Clark County commissioner, a post that he held for ten years, throughout the decade of the 1920s. He continued to divide his business time between Searchlight and Las Vegas, but his success in Las Vegas in a number of ventures—Cadillac dealer, Caterpillar dealer, Boulder Canyon Airways operator, Lake Mead Tour and Excursion Service owner—led him eventually to disengage from Searchlight and concentrate his efforts in Las Vegas. Cashman was locally recognized as the moving force behind many other pioneering charitable undertakings, such as Elks Helldorado and construction of Cashman Field.

Before Cashman leased the phone company, the business had degenerated into an operation that was conducted on the closed-in porch of a home. The water, ice, and electric companies were owned and operated by a Searchlight resident named Walbrecht. With each passing month, the number of customers declined. Most of the businesses tried to maintain their electric service, but by 1915 few homes had electricity, lighted instead by kerosene lamps.[7]

Kerwin recalls that in 1915 there were still a barbershop, a meat market, and seven saloons. The saloons were for men only, with the exception of prostitutes. There were three farms, all on the Colorado River, one on the Arizona side and two on the Nevada side. The transportation of produce from the river was very difficult, and the spoilage was significant; alfalfa was the best cash crop.[8] Freight was no longer carried on the river and Searchlight was not a significant market.

At the beginning of the century Searchlight had had a vast commercial horizon before it, as indicated by significant investments in utilities, transportation facilities, hotels, and stores. But before the end of the first decade, gold waned, and the economic view became more bleak.

12

OTHER CAMPS

Over the years mining activities have taken place in a number of distinct areas surrounding Searchlight. Camp Thurman, a mining district about fifteen miles southeast of Searchlight in the Newberry Mountains, was named for John Thurman, who discovered ore deposits there in 1906. The Newberry range is also home to Spirit Mountain, a unique geological feature that is visible from many miles away because of its distinctive bluish-white color and its height. Its name originated with the Indian belief that the mountain was the dwelling place of departed chieftains.[1] Even today some of the New Agers (a new religion based on nature and ancient Native American traditions) travel to Spirit Mountain and climb to its summit. The trek up this rugged mountain is believed to be difficult enough to bring about a spiritual cleansing. The gesture is deemed to be even more spiritual if the participants are in a state of fasting.

Most of Camp Thurman's activity was conducted by the Lloyd-Searchlight Mining Company. But though this district was always a place of speculation and hope, it never met expectations and produced little gold or other valuable minerals.

The Crescent mining district took its name from the crescent formed by the mountains, opening toward the west, with Crescent Peak near the center. The district, on the northwest side of the peak, had its own post office, which operated from August 1905 until August 1918.[2] Crescent is located about ten miles west of Searchlight, where the New York and McCullough mountain ranges come together. This was one of the areas that was prospected and mined before the discovery of gold in Searchlight.

Indians were the first to discover riches on Crescent Peak, but the natural treasure was turquoise, not gold. The first white man to mine the area was George Simmons. When he developed the property in the early years of the twentieth century, archaeologists from California determined that the mine had previously been worked by Indians. These experts believed that the Indians discovered the claim 200 years before Columbus discovered America.[3]

In October 1903, while visiting the nearby town of Barnwell, Simmons was brutally murdered in an ambush by a former employee whom he had discharged. The attacker was saved from a vigilante hanging and was later tried and acquitted. A legend of lost treasure arose after Simmons was killed. It was well known that he shipped his gems east once each year, and they were allegedly kept in a secret location until transported. They were never found after his death. The legend insists that the gems are still out there, waiting to be rediscovered.[4]

The fact is that Simmons's murder brought an end to the mining of gems at Crescent until the 1970s, when Bozo Quinn resumed explorations. Quinn actually was lured to the area because the demand for turquoise reached a fever pitch in the 1970s. The gems taken then were not of particularly high quality, however, and now turquoise mining has ceased in the Crescent area.

Some gold deposits were discovered about the same time that Simmons was murdered, of sufficient quantity to inspire the rapid rise of another boomtown, called Crescent. This new community had not only its own post office but also its own school and census tract, as well as six saloons, restaurants, a bakery, and a union hall. After the general failure of mining in the southern Nevada area, the 1912 voting list showed only 26 registered voters in the Crescent precinct. The 1910 census showed 66 people living in the mountain town, but the 1920 accounting showed only 26. At its height, in about 1908, Crescent was probably home to about 200 people.

During the last fifteen years, mining activity has been renewed in Crescent. The old gold diggings have been explored by John Yeager Jr., a man with long ties to the Searchlight area who has also earned a master's degree in government from the University of Nevada at Las Vegas. He now is part owner of the old Searchlight Hospital, which has been converted to a single-family residence. Yeager has kept his mining efforts on a small scale but has pursued them steadily. His operation has included the construction of a small ball mill, which is used to mill the ore produced by the mining at Crescent and other ore and tailings hauled from the Searchlight area. The activity has been enough to keep Yeager and a small crew busy for two decades.

One of the early residents of Crescent was Jim Jost; he was also one of the last to leave. When he vacated the area, he literally took the town with him, transporting and then selling all of Crescent's movable buildings to people and businesses in Searchlight. Jost moved the final building in 1920. All that is now left of Crescent's peak years is the concrete foundation of the old post office.

Other than the camp in the Newberry Mountains and at Crescent, mining around Searchlight has been severely limited.The hundreds of mines in the Searchlight mining district, from the Crescent area to the old river, were not community-oriented. They were rarely connected with a permanent residence. The people who mined these disparate claims lived in tents or in other crude encampments. The mines did not usually produce sufficient quantities of minerals to justify more permanent arrangements. There was always hope of new strikes of gold in and around Searchlight, but nothing ever fully materialized.[5]

13

CATTLE AND CROPS

Before the discovery of gold, the Searchlight area was uninhabited, extremely arid and otherwise barren. Ranching had recently commenced in the vicinity and soon would extend into the Searchlight area and beyond.[1] A number of ranching operations merged in 1894 to form the Rock Springs Land and Cattle Company. Its principal brand was "88," and old-timers referred to it as the 88 Outfit. Though the Rock Springs operation did not cover great parcels of grassland like the huge Texas ranches, it was comparably large, at its zenith encompassing 1,600 square miles, or 1,024,000 acres. The ranch covered all of Searchlight, clear over the Crescent range to Kessler Springs, past Piute Springs and then over to Government Holes and the Santa Fe Railroad line on the south. The majority of the ranch's land was actually in

California,[2] but its headquarters were in Manvel, or Barnwell, as it later was known,[3] the railhead for the railroad that served Searchlight.

Not until water was struck in the mines was there any water in the Searchlight area, for surface water was nonexistent. This Rock Springs operation, once established, tried to control all of the area's water sites, no matter how small. Like many cattle operations of the period, it was ruthless in its efforts to maintain control over water. Once the outfit took or bought water rights, others had to go without. As all westerners are aware, water is king in the desert. The Rock Springs company was so important to the people of Searchlight that the *Bulletin* devoted an extremely long piece to the company, with a banner headline that read, DESERT SPRINGS SUPPORT CATTLE. ROCK SPRINGS LAND AND CATTLE CO. HAS LARGE RANGE AND MANY CATTLE. This article explained in detail the scarcity of water and described the vastness of the cattle ranching operation: "It is the largest desert cattle company in the country and its great range which it practically owns by reason of possessing the principal water holes, extends in a general east and west direction from Searchlight to Indian Springs, the latter 10 miles east of Soda Lake, a total distance of 76 miles, and for 70 miles in the general north and south direction from Goffs to the McCullough Springs."[4]

Ranching in the East Mojave is always subject to great variation in rainfall. Ranchers have always shipped cattle in and out in response to the availability of feed, while keeping a base herd of "desert wise" cattle.[5] In spite of the extreme aridity, the range could usually handle as many as 6,000 head of cattle and 100 horses.[6] Some have suggested that the area could handle 10,000 head.[7] Existing records, in fact, show evidence that there were sometimes herds that large.[8]

The owners would work hard to buy or otherwise assume the water that others developed. Their rationale for the takeovers was simple: since the company controlled all the rest of the water, what could anyone else do with such a paltry supply as what remained? The *Bulletin* wrote: "Within a few miles of Searchlight, to the west, there is a very fine grazing country, capable of supporting many hundred head of cattle, which is at present worthless because of the lack of water. . . . Some months ago Mr. Seay discovered water potential. . . . The Rock Springs Land and

Cattle company, always on the lookout for new ranges, is reported to have made a deal with Mr. Seay."[9]

Initially the key water sources in the East Mojave were acquired legally with National Forest scrip. The federal government issued this scrip in exchange for land that it wanted in the newly formed national forests around the country.[10] According to Dennis Casebier, there was no significant controversy between the miners and the ranchers; the real trouble came with the onset of homesteading. Competition then developed between the cowboys/ranchers faction and the farmers/homesteaders faction. Not all historians agree that the Rock Springs Company operated ruthlessly, but almost all agree that when the homesteaders arrived, the cattlemen's tactics became more ruthless, particularly because the homesteaders erected fences that interfered with the free roaming of cattle.

John Reid, my grandfather, at the time a young mining man, made a discovery of water about three miles to the southeast of Searchlight, in one of his mines. Reid said he felt it, tasted it, saw it, and it seemed clean and drinkable. He was confident that the newfound water increased the potential for farming and ranching; the president of the Rock Springs Company, E. G. Greening, inspected the find and made Reid an offer, which he was not happy about but could not refuse.[11] In short, the owners of the company firmly believed that all water within their ranching territory was theirs.

Greening died in the summer of 1910, and his son, Walter, took over the company, which was never as strong after the founding patriarch died. The huge operation went forward, but Searchlight had declined as a good market and homesteaders in the Lanfair Valley were competing for the available land and water. After four years of a disastrous drought that began in 1925, the operation was forced into bankruptcy in 1929.

Hindsight reveals that there are some fairly good grazing lands where the 88 Outfit ruled, the best of them in the hills and mountains where more moisture fell. By far the largest areas controlled by Greening were marginal in production of food for livestock. Ten thousand or even 5,000 head could put a tremendous burden on these arid lands.

The mining camps throughout Nevada, including Searchlight, constituted a market for the cattle and agricultural goods produced by Nevada ranchers and farmers. The hungry miners were good clientele for these new agriculturists.[12] Forty years earlier, during the height of the Comstock, ranching communities had developed in Mason Valley in Lyon County, and in Minden in Douglas County. The Colorado River farms and the cattle ranching around Searchlight served the miners in similar fashion.

The breakup of the 88 began with the sale of the Nevada operation and a small portion of the California property to movie star Rex Bell in 1931. Bell operated the ranch for several years and then divided it, selling part to Al Marshall about 1945 and the rest to Karl Weikel in 1950. The Marshall operation was always marginal, since the land included only a few good water holes. Weikel, however, ran as many as 1,000 head on his ranch, which operated successfully until 1980, when he sold it. Before that, the Weikels sold individual interests in the ranch; for example, Daryl Crow joined the ranch in 1978. Weikel continued to use the Walking Box name, which came from the Rex Bell cattle brand, a box with legs. Weikel's own brand was YKL, a takeoff on his name. The Weikels were heavily involved in Nevada politics, and the old Rex Bell ranch was often the site of Republican Party gatherings. Evelyn Weikel served several terms as the Republican national committeewoman from Nevada, the most prestigious position in the party, reserved for party stalwarts.

West and south of Searchlight is an area that the early settlers called Piute Valley. Since it was not hilly or rocky, some identified it as potentially good farmland; the problem, of course, was the lack of water. To the southwest of Piute Valley is another valley that, astoundingly, lies at an altitude 1,000 feet above that of Searchlight. The highest part of the valley, at 5,000 feet elevation, came to be known as Lanfair Valley. To its east is Piute Springs, and nearby is New York Mountain, so named because the shape of the mountains as seen from a distance resembles the Empire State Building.

Ernest Lanfair, one of the biggest confidence men to exploit the Searchlight region, came to the area that took his name in 1910. Lanfair had previously been a merchant in Searchlight, with a produce company named Lanfair and Wilson. He also owned some mining claims at Camp

Thurman. He built a store and a fake well in the new valley and, through his connections with a California real estate firm, began advertising the agricultural potential of the area. Initially, there was no water in his well, but he installed a windmill to make it look as though there were. When he knew a prospect was coming to investigate the property, Lanfair would transport water from a spring ten miles away and dump it into his fake well, thus fooling the potential buyer into believing that the land had water. The con continued from there; the land he showed to prospective settlers was not his but belonged to the government. Lanfair would charge settlers a fee for instructing them in how to prove up a federal homestead interest.

Even though Lanfair was a fraud, he persuaded Searchlighters to believe that their gold was failing but that a fine agricultural plan was in the process of development.[13] There really was no plan, other than Lanfair's scheme to sell arid, nonfarmable land. In the fall of 1911 it was reported that there were as many as twenty-five different farming operations in Lanfair Valley.[14] None of them proved successful over the long term.

A total about 175 people settled in the valley, nearly all of them as the result of Lanfair's salesmanship, and they made a valiant effort to turn the area into a sustainable agricultural enterprise. Some did fairly well during the rainy years, as the land was fertile. They raised fruit, melons, maize, corn, beans, grain, and even grapes. One black farmer reportedly tried to raise cotton. Though Searchlight was failing and they did not raise substantial crops, they were close enough to the railroad at Barnwell that what they did produce could easily be transported to other markets. Unfortunately, rainy years did not occur often, and dry farming ultimately failed, since it was too expensive to pump water out of the wells. The last Lanfair Valley farm was out of business by 1930.[15] Many photographs remain of the Lanfair agricultural largesse, especially the melon crop, during the wet years.[16]

Dennis Casebier, the preeminent historian of the East Mojave, does not believe that Ernest Lanfair was a fraudulent land speculator or that he practiced deception in his business dealings. Casebier maintains that, in reality, Lanfair fitted the well at his Lanfair facility with a windmill in 1914 and that it even to this day produces water. There is much evidence to the contrary, however, regarding the ethics of Lanfair's business practices. It can be argued, though, that the blame for the agriculture debacle does not rest with Lanfair alone, that the U.S. government shares responsibility since it allowed the area to be homesteaded.

Early on, the Colorado River seemed to promise a limitless agricultural bonanza. With the river's constant supply of water and the passable soils, which were rejuvenated when floods occurred, the area held real agricultural potential. When the railroad came, bringing access to supplies and a means of shipping farm produce out of the area, the possibilities for farming became attractive. Several operations survived for fifty years, until the waters of the Davis Dam covered the properties. But the river farms were always marginal, mainly because of the borderline soil and the difficulty of transporting the produce once it was harvested.

With the development of Searchlight came the maturation of the agricultural interests along the river. Riverside Ranch was developed right on the banks of the river, as were virtually all the other ranches east of town. The Riverside had 100 head of cattle and 300 acres of irrigated land, which included alfalfa and vegetable crops. The products were marketed and sold in Searchlight, Needles, and other camps along the trade route.

By 1905 the Hutton River Ranch was made a working property, consisting of 320 acres of bottomland along the river, with another 100 acres cultivated off the river.[17] The success of the Hutton River Ranch reportedly resulted from Hutton's invention of a water wheel that moved water from the river to a reservoir he had constructed, allowing him to irrigate. This ranch, with 160 acres acquired under the Desert Land Act and another 160 acres acquired by homestead, was sold to two men named Gerhart and Holton in 1905. It was the only property available along the river, near Searchlight, for any government entry; the rest of the land was unsurveyed. Initially the tract was believed to be suitable for alfalfa, with as many as seven cuttings a season; it also supported plantings of sweet potatoes, grapes, and hundreds of fruit trees.[18] In the summer of 1909 came reports that the Weaver Ranch on the Colorado River had a fine crop of fine melons, which sold for three and a half cents a pound.

That same summer, two acres of the ranch's prime farmland were swept into the raging Colorado.[19] This was the only reported loss, but it was likely a common occurrence along the treacherous river. Normally the river was narrow, but when the winter snowpack melted, the river widened to many times its normal size and wiped out everything in its path. At most times, the river water was saturated with sand. To get usable water from the river, one would have to draw a container of water and let it sit long enough for the sand to settle to the bottom. During a flood season, one would wind up with nearly half a cup of sand to go with half a cup of water.

As the years passed, the Depression affected the area, and the construction of dams upriver curtailed freight traffic on the river. The individual ranches were consolidated to form one operation. The last ranch on the river below Searchlight was known as the Verzanni property, named after the operator. Ernest Verzanni had eleven children, and his ranch supported as many as 600 head of cattle. Supplies were delivered once each month from Las Vegas. The landmark of the Verzanni Ranch was an immense white house built for the very large family. Even as late as World War II, interesting events occurred at the ranch. Along the river near the ranch, the U.S. Army conducted significant military activity. The Verzanni children remember the American soldiers' giving them leftover food and once treating one of the children for a burn. Even though it was likely not factual, the Verzanni children believed that a prisoner of war camp was located nearby. When the floodwaters of the new Lake Mohave, created by the construction of Davis Dam, wiped out the big white house and the rest of the Verzanni Ranch, the family received no compensation.[20] Today, if the government wiped out a person's livelihood, the damaged party would be compensated for the loss.

Though Searchlight was too arid for substantial agriculture production, people saw that the area just west of the town was flat and had good soil, and they continually tried to grow crops. It seems that they ignored the obvious.

Soon after the Rock Springs Company convinced John Reid to sell his water rights, he tried farming again. This time he fenced forty acres of his homestead on the flat west side of town. He moved his house and barn to the property, planning to grow vegetables irrigated with well water pumped up by a windmill. Like similar endeavors, the plan ultimately proved economically unfeasible.[21]

As early as 1905, there was a movement to use the town's garbage to feed chickens, an early recycling project. It was believed that the local market would easily buy all the eggs, at fifty cents a dozen, as well as all the chicken meat the ranch could produce.[22] No reports as to the success or failure of the proposed chicken ranch have survived.

At a couple of the local mines where dewatering operations were employed in an effort to get at the ore, there was some success in growing produce. At the M&M, just north of town, huge squash (one of them weighing more than four pounds) were grown at the mine and delivered to the people of Searchlight, reminding one resident of the Iowa farm where he had lived.[23]

Benjamin Macready made an additional contribution to the life of Searchlight as a farmer. In the summer of 1912 he and his son, John, grew fruit and vegetables, especially peas, at the Santa Fe Mine, which Benjamin was operating—the same mine that struck water at less than 100 feet and sometimes supplied the whole town.

At one time Searchlight also had its own dairy. The barns and corrals sheltered eight milk cows, and milk was delivered to the doorsteps of Searchlight residents for twenty cents a quart, less than they had to pay for canned milk. The operation was managed by a dairy farmer named Hotchkiss, from Eau Claire, Wisconsin. There was hope of also having a poultry farm at the same location.[24]

Clearly, Searchlight's record in supporting vegetation is less than spectacular. From 1920 to 1970, it was rare to see a tree in the area. The townsite itself has extremely rocky ground, and the outlying areas, where the land is fertile, always suffer from a shortage of water. The poor conditions, however, did not deter promotion and speculation for the business of agriculture. As Searchlight's mining operations were declining, one big promotion proposed to sink a series of wells to irrigate 20,000 acres on the level land just west of the town.[25] The plan fizzled. Another promoter envisioned as many as 5,000 cattle grazing within view of Searchlight.[26]

A number of interesting cowboys ran small operations in and around Searchlight. Gene "Sharkey" Myers, the deputy sheriff following Big John Silveria, ran more than 100 head of cattle, principally in the Spirit Mountain area. He also was a prominent rodeo cowboy. E. B. Davis ran cattle northeast of Searchlight and was a fine rodeo roper. His daughter, Nola Jean, was crowned Miss Helldorado Rodeo in 1957. She was judged on her horsemanship, appearance, and personality. Helldorado was for decades the premier Las Vegas tourist event. It was based on a western American theme and consisted of a week of rodeos, parades, a carnival, displays, barbecues, and other activities. Another character was Joe Kennedy, who ran a few head of cattle of his own and worked cattle for others. He was known for his big wad of chewing tobacco, and when asked why he always carried his .44-caliber pistol, he would casually reply, "They didn't make a forty-five."

Cattle ranching came to an end in the Searchlight area as a result of the Endangered Species Act (ESA), which identified the desert tortoise as an endangered species. All cattle were removed from about 440,000 acres, and Searchlight was deemed to be an appropriate place to allow the tortoise to reestablish. The designation of Searchlight as the tortoise's habitat allowed building construction to resume in the Las Vegas area, where previously the U.S. Fish and Wildlife Service, which administers the ESA, had imposed a moratorium on construction because of the tortoise population.

14

COMMERCIAL COMPETITION

People who lived in or passed through Searchlight after 1920 would have found it difficult to believe that this town was at one time one of the most modern in the state.

The local newspaper, in 1906, less than ten years after Searchlight had been formed, vividly described its rapid growth: "Returning old timers standing at the foot of Hobson and looking up the street rub their eyes in bewilderment. And well they may; for great have been the changes during the past year. Where formerly stood a few scattered tents and one or two weather worn frame buildings now looms a long street lined on either side with substantial buildings painted and of pleasing appearance. Today the center of the street is torn up from end to end while a six-inch water main is being laid. How different from days not so very long ago, when not at any one time was there sufficient water in the entire camp to have filled a single length of the now long line of pipe. At the twilights hour arc lamps illuminated the street, and the different places of business are aglow with incandescent lights. In the broad, well-kept street of today there is no brush, no rocks or no maddening cacti; and belated pedestrians serenely amble homeward beneath the glare of the electric lights."[1]

The business community developed significantly despite the fact that the Quartette Mine had erected its own company town, with a store, bunkhouses for workers, and homes for key employees. The mine had its own musical band, baseball team, and tennis court. A few of the other mines had a home for the boss, but none had all the appurtenances of the Quartette.

A review of the advertising section of the local newspapers brought to mind a long-established eastern city of the day. Whether it was a beauty salon or a barbershop, Searchlight had every modern convenience.

When the Santa Fe Railroad attempted to establish a new townsite, businessmen in the community objected and even published a petition in the local newspaper announcing their concern. The state of early trade and commerce can be gleaned from a review of the business names that appeared in that petition: Searchlight Lumber Co., Searchlight Bank and Trust, Searchlight Hotel, Searchlight Supply, Searchlight Stage and Freight Co., California Saloon, French Restaurant, Burdick and Thurman Liquor, Pioneer Barber Shop, Gem Saloon, Searchlight Drug and Mercantile, Col. McCrea Mining Supply, Dr. Jensen M.D., Thomas General Merchandise, Buckler General Mercantile, Brown and Gosney Co., Godin Dry Goods, Albright Contracting, Wheatly General Merchandise and Hotel, Searchlight Publishing Co., Searchlight Development Co., Buckmaster Horse Lodging, Star Hotel, Riverside Ranching Co., Jones Real Estate, Moore and Smith Hardware & Plumbing, Woodins' Mens Clothing, Durdel Surveying, and L. L. French, Esq.[2] The petition listing provides a sense of the variety of businesses, but the establishments whose names appeared there constituted only a fraction of the full community of commerce of Searchlight.

One of the businesses that advertised frequently was the Searchlight Dairy, owned by R. J. Oppeduk, who promoted daily delivery of fresh milk. He even had a phone number that a customer could call for service, a rare feature anywhere in Nevada at that time.

Not only was there a multitude of bars in the mining camp, but one could browse the paper and see an advertisement for Dora Boyles, an instructor in voice and piano who even listed her training and education at Yankton College Conservatory. On the same page appeared an advertisement for the services of Bertha Polley, a public stenographer.[3]

The town also had its share of merchandisers of building materials. There were at least two large lumber companies. Searchlight Redwood Lumber Company ran a one-third-page ad touting the best redwood lumber and other products for building in the desert, and Searchlight Lumber Company also dealt in mining timbers, emphasizing its ability to deliver carload lots of lumber and other products. The Moore and Smith Company claimed to be fully equipped to perform all kinds of general repair work, even the construction of corrugated iron tanks and air pipes.

By 1906 the people of Searchlight had their own hospital, a structure of twenty-eight feet by thirty-eight feet that could accommodate ten patients. Initially called the Mine Operators Hospital, it had a nurses' room, an operating room, and a dispensary.[4]

There were two physicians who advertised their services. Dr. F. J. Nutting, physician and surgeon, indicated that he had a full line of drugs, medicines, and sundries, and—equally important—current magazines, plus the finest brands of Havana cigars.[5] His competition was the understated Dr. C. A. Jenson, who only sought business as a physician and surgeon. It is amazing that Jenson got any business, since another doctor in town, Dr. W. Leroy Fuller, advertised that he was equipped to handle all cases, and, because he advertised his own Searchlight Hospital, he gave specialized treatment to those who were confined. It is believed that this hospital was the same as the Mine Operators Hospital.

There were several lawyers who actually practiced in Searchlight and a few more who advertised their services but whose offices were in Los Angeles. Some of the personals in the paper offered enticing reasons for frequenting a particular business, as well as illustrating the quality of life in a booming mining town. For example, "Mr. J. C. Walter wishes to inform the public that he has added another chair to the Pioneer Barber Shop, which is in charge of Mr. E. J. Thompson, an experienced tonsorial artist, who formerly operated in the best Los Angeles shops."[6] Or, the most enticing advertising, covering many different goods and services: READY TO WEAR = TRIMMED STREET HATS, Mrs. Byrd, and L. Lewis EXPERT CLOTHES CUTTER, and WOODWORTH WOOD PRODUCTS, and THE GEM SALOON, also known as THE THIRST PARLOR.

Even though this small community had a fair number of professionals, they did not appear to be involved in the politics of the town. In fact, until the town's mines failed, there was really no attention to governance, with the exception of concern about the town's school. During the first ten years of Searchlight's existence, no elections were held.

The commercial enterprises were unique, and many of the ads for them—such as A NEW BAKERY NEXT TO THE CHINESE LAUNDRY—revealed the variety in early Searchlight. There were paint stores, a steam laundry, an ice cream parlor that included a soda water fountain, the Palace Grill (which bragged of serving meals at all hours), and a butcher shop featuring beef from the Crescent range.

In addition to hotel dining rooms, the town had several specialty restaurants, such as Millen's Spanish Restaurant and a restaurant offering French cuisine.

The prices at the local stores were clearly displayed. On April 9, 1909, one could buy a two-and-a-half-pound can of tomatoes for ten cents, or corn for the same price, but the same size can of peaches cost twenty cents. One could purchase a high-grade work shirt for fifty cents and a pair of coveralls for eighty-five cents, but a good hat cost $3.50.

In Ely at this time, canned produce was somewhat higher than in Searchlight. Peaches cost an additional nickel per can. One could buy ten cans of tomatoes in Searchlight for what it cost to buy six in Tonopah. A good Stetson hat in Tonopah cost the same as in Searchlight.[7]

From the early days of discovery, when the parched land had little to offer its inhabitants, until the glory days of the Quartette Mine, the business community performed at maximum efficiency. The town's enterprises, in the first twenty years of existence, were always on the cutting edge of modernity. Only the best and most advanced technology was present.

Not only did Searchlight have a telegraph, which was fairly common, but the townspeople quickly craved a telephone system, as was first reported in September 1905, when the Searchlight Telephone, Telegraph, and Electric Company (STTEC) was incorporated to establish a local telephone exchange and a long-distance system. Initial arrangements called for the installation of forty telephones in Searchlight.[8]

The system was complicated because two companies were competing for the Searchlight business. The Searchlight-Manvel Telephone Company, which was principally owned by the Quartette Company, wanted to extend the long-distance connection to both Nipton and Manvel.[9]

The independent STTEC had the advantage, because it had already signed contracts with the Northwestern Power and Telephone Company by the time Searchlight-Manvel showed interest. This was important because the Northwestern Company had exclusive rights to the long-distance line between Los Angeles and Salt Lake City. STTEC also had a signed contract with Western Union to handle the technical aspects of the tie-in with the Los Angeles and Salt Lake facilities. Upon the successful completion of these arrangements, including the signing of the contracts, STTEC began construction of the plant and ordered poles and other equipment.[10]

By February 1906 all of the equipment was in place, and the phones were connected with the town and the mines. The next project was the construction of the line to Nipton so that the demands for long-distance service could be met.[11] The construction project to Nipton was not exceedingly difficult, since the terrain over the twenty-two-mile distance inclines gradually to the west of Searchlight for about eleven miles, to the Crescent area at an elevation of about 6,000 feet, and then runs downhill for the remaining ten or eleven miles. The construction of the line, though, went slowly, holding up the new business, since by May the company reported having not forty but eighty poles in operation, and the demand for new telephones was heavy. The line was, however, reported completed by the first part of June.[12]

Alexander Graham Bell invented the telephone in 1876, and the first exchange was opened in Connecticut in 1878. Service did not extend across the American continent until 1915, so it is fair to say that early Searchlight was up with the times, a very modern little city.

Searchlight was also at the forefront in bringing another modern convenience to its residents—electricity. Even though electricity had been discovered in the eighteenth century, it was really not developed until the end of the nineteenth century. So when in late 1905 plans were announced by Western Telephone Company, a new business in town, for an electric light and power system, the town was understandably pleased with itself. In March 1906 it was announced that in order for the project to be completed, subscribers for 1,000 electric lights were needed.[13] Notwithstanding the difficulty of attracting the required customer base, the company declared its plans to develop an electric power plant.[14] The company reported in June that the demand for electricity was even greater than anticipated, and it ordered the most modern equipment available, including a 120-horsepower gasoline engine with one flywheel placed in the center between two horizontal cylinders. The advantage of this engine was that it could continue to operate on one cylinder if the other broke down. The equipment, manufactured by General Electric, would furnish up to 2,000 lights and supply power to the mines.[15] Even though the initial demand for power was strong, the final number of subscribers was lower than anticipated. By September only 500 had applied, and only one mine was on line. The company was betting that once the line became operable more customers would sign up.[16]

Getting the electricity to town proved much more difficult than anticipated because of a strange sequence of mishaps that caused awkward delays in delivery of the generating equipment. Materials were ordered from General Electric in June, but a train wreck damaged the goods. A duplicate shipment was sent via Oakland, but it was confiscated and used as emergency earthquake relief supplies. A third shipment got as far as Los Angeles and then just disappeared. It was lost for weeks, then finally reappeared in November and was shipped to Searchlight.[17]

All the equipment was tested during November, and the power company announced that the system worked "beautifully." At this time only

400 lights were on line, with power available from sundown to midnight. There was hope that in the future electricity would be available for the entire twenty-four hours of the day.[18]

Again, Benjamin Macready was involved in progress in Searchlight, this time as one of two partners in the establishment of electricity in the town. The plant was built at the Santa Fe, a mine he controlled. The energy produced was 2,300 volts, reduced to 120 volts through a series of ten transformers. The poles were thirty feet high and were set back eight feet from the street as a safety factor.[19]

On November 23, 1906, the headline in the local paper read: ELECTRIC LIGHT PLANT IS A COMPLETE SUCCESS. Everybody was so bullish on electricity that the town talk prophesied that the plant would soon carry power to Eldorado Canyon, Crescent, and the tributary camps to alleviate the need for the mines to have their own power plants.[20]

Another achievement that would ensure the town's commercial viability was a successful solution to the area's water problems. Water was scarce and difficult to access. Many stories from the camp's early days report the lack of water and the suffering and, occasionally, death of travelers in the area. The river was fourteen miles away, Piute Springs at least twenty miles, Grapevine Canyon Springs nearly twenty miles, Summit Springs about three miles. A series of small springs at Spirit Mountain and a few others intermittently spaced throughout the McCullough Range provided water for individual mines throughout the area, while Searchlight got its own water from mines within Searchlight itself.

The earliest writings in Searchlight recognized the need for water: "We have several hundred square miles included in this mineral belt from all points of which valuable discoveries are announced. Throughout this large area natural springs occur at intervals of from 15 to 30 miles, giving a moderate supply of water the greater part of the year. A limited amount of work on these springs would furnish an ample and permanent supply the year round, under all climatic conditions; but any point 5 miles or over from water supply is practically barred from the prospector, except to run over it in a most haphazard manner thus affording only very remote chances for new discoveries."[21] It was because of the belief that there was no water in the entire area of a sufficient quantity to operate a mill that the Quartette Company built its mill at the river.

In spite of the critical shortage of natural water, once the mines were started, surprisingly large quantities of water were found. The Quartette spent a small fortune building the mill and the railroad to the river and then immediately learned that the mine's shaft had large amounts of water, so much in fact that the excess was allowed to run down the washes below the mine.[22] In Searchlight, as in most other mining towns, the domestic water came from the mines. (Ely proved to be different; there, water for the mill and for the town came from a ranch that the mine owners purchased.[23])

The Quartette was not the only mine to strike water. The shallowest strike in the district was at the Santa Fe, at 168 feet in the incline shaft but only 90 feet vertically.[24] The Quartette's water came at 500 feet. The Good Hope hit vast quantities of water at 300 feet. The Duplex, when shut down, quickly developed 350,000 gallons of water in its 300-foot shaft. The Southern Nevada Mine hit water at 225 feet, with an average of 25,000 gallons a day. Some of this water was used in the mine's stamp mill, but most of it washed down the way from the mill and mine. The Pompeii hit so much water at 262 feet that the operation had to be shut down until methods of removing it could be determined.[25] Much of the water in other mines could be put to no practical use.

Since no one in Searchlight knew that there was water in the mines until 1903, various other options for supplying water to the town were explored before that time, but with little success. The developers of water in Searchlight started at the beginning of 1903 and would have been able to establish a water company but for the union strike that hit Searchlight so hard.[26] The company, Searchlight Development, chose the Santa Fe as the location of the town well. By February the main water lines, consisting of 2,000 feet of pipe, had been laid and a 15,000-gallon storage tank put in place, as well as ten fire hydrants. Customers had to purchase water meters at a cost of $10.40 each, and they had to make their own connections from the center of the street, with the pipe being furnished by the company that guaranteed the meters.[27]

Improvements to the well and its carrying capacity were an ongoing effort. As the system got bigger, several different storage tanks were used, one for filtering and the others as holding tanks. The full capacity, utilizing all the tanks, reached 85,000 gallons.[28]

The water supply was filtered to remove silicate of lime, which though not harmful, gave a sandy quality to the water. The lime element was removed by percolating water up through one tank and then down through another.[29] Filtered water was a luxury not enjoyed by other Nevadans.

The success of the water company in Nevada was clearly illustrated by the development of a fire department, which was possible because almost two miles of pipe started at the lower end of Hobson Street and extended to the head of the avenue, with appropriate arterioles. With a fire department, it would be easier to obtain insurance for the business district. In the spring of 1907 an impressive demonstration showed the power of the water system. It took two men to hold each of the two powerful fire hoses, which were three inches in diameter and shot water 100 feet into the air. So great was the pressure that the men had to aim the water stream away from the buildings for fear of causing structural damage. Even with such pressure, it was estimated that the various tanks held enough water to keep the hoses going for up to eight hours.[30]

It proved difficult to raise the money for the fire department and eventually took a year and a half to get the department and the fire hydrants in place.[31] Even though the fire-fighting equipment did show some promise of being able to stop the spread of fires in the early days, over the years several significant fires claimed many of the town's permanent wooden structures. As in most pioneer western mining towns, Searchlight's history is traced by the dates of the various fires.

By 1905 the Searchlight Development Company had also begun work on an ice plant. The equipment was to be capable of making five tons per day; the full capacity was never to be used, however, for the necessary cold storage space was not available.[32] It was only on July 5, 1905, that the first cake of ice was pulled from the plant. The product was the best, having been filtered twice, the power being furnished by a 16-horsepower gas engine. The ice was used primarily for businesses and mines. None was exported, since no feasible means of transportation became available until after the town and its major mines had already begun to decline. Ice was a big export in Las Vegas, but that city had access to the railroad to ship the ice to the mining camps in Nye County.

The residents of Searchlight tried their best to create a totally modern city, but the elements, the location, and the lack of sustainable mineral resources doomed the worthwhile project to failure.

15

POLITICS

In 1902 an unknown young man by the name of Henry Hudson Lee attended the Nevada State Democratic Convention, held in Pioche. With no intention of becoming a candidate, the young man surprised his family, friends, and the whole county by allowing himself to be persuaded to run for Lincoln County recorder and auditor. His chances of winning against Harry Turner, the incumbent, were deemed to be quite remote, since Turner was regarded as politically prominent and well known in the county.

Lee won an upset victory, however, and believed his win was attributable to the voters of Searchlight: "I had the best of Harry Turner because he hadn't been in this part of the state since Searchlight had been active. I had a big advantage over him because I contacted a lot of people in this part of the county that he had never seen, or even heard of."[1]

Henry Lee was the first to learn that Searchlight was a power in county and state politics; in the future many would follow his example. In the next few years a steady stream of county and statewide politicians traveled to heavily populated Searchlight. The trend continued in later years, even though the population of the camp plummeted in the early months of 1908.

Newspaper headlines announced that Governor John Sparks and the whole Democratic ticket were due to campaign in Searchlight before the November 1906 elections. This attention from candidates continued into the next decade, even as the census dropped. In the November 1910 elections the Republicans came to town with Tasker Oddie, of Tonopah, who was running for the U.S. House of Representatives and who subsequently became governor of Nevada and a U.S. senator. He was accompanied by George Nixon of Winnemucca, soon to become a U.S. senator. They came to town with the whole Republican entourage, even though Searchlight was many miles from their homes. Only a week later, the Democrats arrived, bringing with them future governor Denver Dickerson.[2]

Some of the dialogue of the early days of Searchlight could fit in just as well as political banter of today. In 1904 the *Searchlight* wrote of the difficulty of finding a candidate for the U.S. Senate seat. At the time, state legislatures chose the U.S. senators. Governor Sparks said he didn't want the job because "Senators are elected by how much money they have and how lavishly they spend it." This same statement would stand today amid the clamor about spending in elective politics.

Colonel Hopkins, of Quartette fame, gave the same reason for not seeking the job of U.S. senator, adding that he considered a purchased seat in the Senate to be a disgrace rather than an honor.[3] Even though U.S. senators were appointed by the legislatures, the campaigning was deliberate and very expensive.

A humorous and invective-laden opinion published in the newspaper a couple of years later shows that cynical attitudes toward politicians were prevalent even then: "Will someone explain just why it is that many candidates invariably appear on the platform dressed more poorly than any one else in the audience. The presumption is that the speaker is endeavoring to disguise himself as a horny handed son of toil and thus make a hit with the working man. However, the working man like every other intelligent voter, figures that an office seeker should be one who is strong in the head rather than in the back, and further that he should at least have been successful in his profession to the extent of being able to appear well and suitably dressed."[4] Political posturing seems not to have changed these past ninety years, as candidates are still trying to appear less elitist by wearing plaid shirts, sweatshirts, and open collars.

What was entirely unlike modern political commentary was a short piece in the *Searchlight* reporting that all Republicans and most Democrats were satisfied with the work of the Fifty-ninth Congress.[5]

To demonstrate the political power of early Searchlight, one need only look at the Democratic county convention of 1906, to which Searchlight sent seven delegates, Las Vegas five, Caliente six, and Crescent and Nelson one each.[6] This period would be the apex of Searchlight's political history.

With their increased economic and political power, the residents of Searchlight wanted the county seat to be closer than 250 miles away in Pioche. Government services to their booming town were weak, at best, and response from the county was slow. In the view of Searchlight's citizens, Lincoln County's northern communities were envious of Searchlight's success and too heavily represented in the affairs of the vast county. Las Vegas residents felt just as strongly that either the county seat had to be changed or a new county formed.

Recorded complaints were registered as early as 1904. The Searchlight newspaper reminded residents that they should unite behind a southern Nevada candidate. The paper became even more specific in 1906, noting that while Searchlight had the most voters and paid the most taxes of any community in Lincoln County, it received nothing in return. It was argued that it was a waste of time to complain to county officials, as they didn't care about Searchlight.[7]

U.S. senator William Stewart introduced a bill in the Nevada legislature creating a new county in southern Nevada, to be called Bullfrog. It must be remembered that at the time, U.S. senators were chosen by the state legislature and therefore had more contact with the legislators than do senators in the modern era, who are elected. Searchlight didn't like this new county idea because the county seat would still be too far north, but the idea did stimulate discussion as to whether the sprawling county should be divided and the county seat moved to Searchlight, Las Vegas, or Caliente from its current location in Pioche.[8]

The general assumption was that everyone except Pioche residents favored removal. Surprisingly, however, the people of Pioche strongly preferred the division, which would allow them to retain their county seat regardless of a split or other partition. The *Bulletin* opined that the county seat should move to a more central location and that Las Vegas would be the best bet, leaving the county to split later. At this time, there was no thought that a new county would be created. It was generally agreed that if the decision were simply to choose a new county seat, Las Vegas would be the logical choice because it was more centrally located than Searchlight. The idea of moving the county seat failed, however, and efforts shifted toward the formation of a new county.

In August 1908 a Lincoln County Division Committee from Carson City came to Searchlight to explain why Las Vegas should be the county seat. The presentation laid out many enticements, especially that the city of Las Vegas would furnish free quarters to county officials for three years and would apportion the county debt based on assessed valuation of privately held property. It would take three-fifths of the qualified voters to sign a petition, but they would also have to be taxpayers.

This created many problems, since 60 percent of the taxable property was owned by the Salt Lake Railroad and the railroad was not a qualified taxpayer. So it was not feasible to distribute a petition. For that legal reason, division, rather than removal, was the only solution for the south. It also helped the proposition that William Clark, president of the railroad, was supporting the county division. The *Bulletin* reported that the people of Searchlight were not enthusiastic either way.[9]

A county division bill was again introduced in the Nevada legislature in 1909, as in earlier sessions. During the hearing stage of the bill, the railroad announced that it was going to move its repair shops to Las Vegas. This was an important step because it would increase property values and population in Las Vegas. Searchlight was already losing the population contest with Las Vegas, and selecting Searchlight as the county seat was now out of the question. Had the division taken place in 1905 or even during the 1907 legislative session, Searchlight could have prevailed. The division bill passed both houses of the legislature and was signed by Governor Dickerson on February 5, becoming effective on July 1, 1909.[10] The county was named after William Andrews Clark, a mining baron and former U.S. senator from Montana, whose claim to fame in Nevada was that he built the railroad from Los Angeles to Salt Lake City, which was the main factor in locating the county seat in Las Vegas.

Bullfrog County became a hot political issue in the 1988 U.S. Senate race between incumbent Senator Chic Hecht and Governor Richard Bryan. Hecht alleged Bryan was soft on the issue of nuclear waste storage in Nevada because, as governor, he had signed a bill in 1987 conditionally creating Bullfrog County. The allegation was that this legislation was intended to appease the nuclear power industry. Bryan won the election contest by a good margin and became known as the most outspoken foe of the storage of nuclear waste in Nevada.

In 1907 the Searchlight Chamber of Commerce investigated the feasibility of incorporating the town. The only opposition came from the proprietors of the town's saloons, who feared excessive taxation. Searchlight became an incorporated city in October 1907, and the first municipal election was held on November 26.[11] The new council consisted of Mayor Emerson with council members Light Wheatly, H. A. Walbrecht, and a Mr. Beattie. Emerson is the only mayor Searchlight has ever had. By 1920 the city no longer elected officers, depending instead on county government, as it still does today. In the past twenty years Searchlight has functioned with a town board, popularly elected by Searchlight residents and then confirmed by the Clark County commissioners. Today's structure is a far cry from the time when the town was the premier candidate for county seat.

Over the years Searchlight has provided several notable Nevadans who served in elective political office. James Cashman, the early Searchlight businessman, served as a member of the Clark County Commission.

Dick Arnold was born in Virginia in 1989 and first appeared in Nevada as a twenty-year-old in Moapa Valley, where he worked in the cantaloupe farms. He moved in 1924 to Searchlight and worked as an inspector at the California-Nevada border during the epidemic of hoof-and-mouth disease among the area's cattle.

After the death of his wife in 1925, Arnold married Frances Knowles, the daughter of Ella Knowles, Searchlight's most successful merchant. It was from Searchlight that he was elected to the Nevada Assembly in 1934, and in 1938 he won a seat in the Nevada state senate, a significant position since at this time each county was represented by a single senator. Even though Arnold was only forty-five years old, his health began to fail rapidly. Before the 1939 legislative session convened, Senator Arnold was severely burned. He resigned from the legislature and died in August of that year.

This bespectacled young man would have grown to be a significant power in Nevada politics had he not died so young. A Las Vegas newspaper said it best: "He was a man to whom all who knew him became strongly attached because of his loyal and kindly disposition. He was always active and full of ambition for his chosen community and performed many acts of service."[12]

Rex Bell, another Searchlight resident, was twice elected lieutenant governor of Nevada, in 1954 and 1958. Clearly, he would have been elected governor had he not died of a heart attack during the 1962 campaign. His son, Rex Bell Jr., served as a justice of the peace in Las Vegas and as a distinguished district attorney of Clark County from 1986 to 1994.

Stan Colton, the son of Gordon Colton (the grandson of the founder of Searchlight), was elected state treasurer in 1978 and subsequently served in the appointed post of registrar of voters of Clark County.

16

SCHOOLS

About the same time that Searchlight was booming, Tonopah, Goldfield, and Rhyolite were also maturing. A look at the formation of schools in these communities is instructional. At Tonopah, as early as 1901, almost immediately following the discovery of gold in the area, a mass meeting was held at which citizens demanded schooling for the children. An appointed committee recommended that all citizens be assessed a tax for the construction of a school building and for the funds necessary to employ a teacher. A teacher was hired for $90 a month, a very decent wage for the day.

The county commissioners almost immediately formed a school district for the burgeoning camp of Goldfield, and by 1904 it also had a teacher.

In Rhyolite, citizens also demanded a school, even though there were only five children in town and only three of those were of school age. When the commissioners refused to act, the townspeople held bazaars, dances, and other social events, and by 1905 they were able to finance their own school.[1]

Searchlight's experience in establishing a school was quite similar to those of Tonopah, Rhyolite, and Goldfield. We have no account of the first school in Searchlight, but we know the town had a school very early in its history. From the first articles about the school in the town's newspaper, we know that a school existed before 1903 and that by that year Searchlight had its own school building.

At each session of the Nevada legislature, which meets every two years, the superintendent of schools makes a report on all Nevada elementary and secondary schools. Searchlight is first mentioned in the reports for the years 1899–1900. In 1900 the teacher in Searchlight was Belle Van Dunder, and she was paid $65 per month.

By 1903 there was a school board, which remained in existence until 1956, when the different Clark County school districts consolidated into one district. The Searchlight School District was in district number five of the Nevada State Department of Public Instruction and, therefore, reported to the State Board of Public Instruction, which had oversight responsibility. The first mention of the schools in the local newspapers appeared in the fall of 1903 when a new teacher, Miss Alma Tuttle of Perris, California, was hired because she was not only an accomplished teacher but also an elocutionist and a musician. The superintendent's report shows that Laura Daugherty replaced Belle Van Dunder at the same salary. During the period 1900 to 1904, when Miss Tuttle started teaching at $75 per month, town leaders filled the position of clerk of the local school board. F. W. Dunn, the man who stimulated Hopkins's interest in Searchlight, was succeeded by the publisher of the *Bulletin,* Howard Perkins. School was ordered to begin on October 15. Here the board indicated that it would be using the old school (the first reference to a school building that already existed), even though they preferred Miners Union Hall, which couldn't be used because of the ongoing strike.

In the fall of the next year, a new school board agreed to rehire Miss

Tuttle, as she had performed well in her duties the previous year. The board also resolved to seek funds for construction of a new school building, having already decided the preceding March to raise $600 for a building fund by assessing a special tax.

As would be expected, the struggles of the school and its teachers were exacerbated by the dispute between union and nonunion workers, and between the Quartette Company and the townspeople. One battle developed over whether the new school should be located on Quartette property, where most of the people were employed, or on other property closer to the center of town. Another spat that made the paper had to do with who would pay for the annual Christmas party and where it would be held. The best place to hold it would have been in Miners Union Hall, but the nonunion parents rebelled. It was also thought improper to have the Quartette pay for the annual event. Finally, an anonymous donor who was not identified as siding with either faction gave money for the party. This incident shows the patent hostility between the town and the camp's largest employer. Though no one really knows, it seems likely that the anonymous donor was the Quartette Company.

The battle continued in the 1904 school board elections over the possibility that board members might experience personal gain by virtue of the site selection.[2] The school was built in Searchlight and not on the mine property, but later it was again moved. There is an oral tradition in Searchlight asserting that during the height of prostitution in town (which followed the decline in mining activity), a dispute arose because, according to state statutes, the school was located too close to a saloon where prostitution went on. To solve the problem, the bar remained where it was and the school was moved, thus eliminating the legal impediment to prostitution. While the story likely is factual, there is no independent proof, as the incident occurred when there were no newspapers, and there is no living witness to confirm this 1920s tale.

In 1906 the school board again decided to hold a special election, this time to raise $1,500 for school expenses, including hiring a second teacher. The school needed another room, and it was decided that the school year should be extended to eight months. The problem was intense because 90 percent of all the school taxes (these being property taxes) were paid by the mining companies. For the last assessment period the Quartette argued that it had overpaid its assessment; the company took the issue to court, where it prevailed. This decision created the need for another levy, which had to be approved by the voters themselves.[3]

The papers also periodically published the number of students and their names. The most ever reported in either the *Searchlight* or the *Bulletin* was thirty-nine students, in 1907. (At this writing, some eighty-eight years later, there are thirty-nine elementary school students.) The paper also printed the grades of all the pupils, whether good or bad. By 1907 the Searchlight school had two teachers, sisters Anna and Rose Falconer, who were each paid $80 per month. By this time an addition to the school had been built.

The Crescent mining area also had an active school operation, as indicated by the fact that it had twelve students in school in 1907. This statistic is perhaps the best independent proof of the size of the operations at the Crescent.

Most of the boys and young men in the mining camps were not interested in schooling. Because of their reluctance, the school board published a Nevada statute that ordered every child between the ages of eight and fourteen who lived within two miles of a school to attend classes.[4] The law was specific in requiring that these children attend school for at least sixteen weeks each year, at least eight of which had to be consecutive unless they lived more than two miles from the school.[5] In Searchlight, this law was rarely enforced, and many of the town's children went unschooled.

After the mine leasing period gained momentum, about 1910, the school population dropped precipitously. In 1912 the school opened with eleven students, in all grades one through eight. In the 1913 school year there were also eleven students, but only seven of the eight grades had students.[6]

The school building on top of the hill in the middle of town was replaced in 1942 with a cinder-block building that had two full rooms and two indoor toilets. The old wooden frame school was moved to property near the Needles Highway and used as a bar and casino. Shortly after the move it burned. The new facility had an oil heater, which replaced the wood and coal stove of the early twentieth-century building. This 1942 school remained until 1992, when a new building was constructed at an altogether different site on the road to the lake, next to the community center.

The history of the Searchlight schools after 1912 can best be reconstructed by reading Arda Haenszel's *Searchlight Remembered* and from conversations with students from that time. The actual school records no longer exist, in either the Clark County School District or the Lincoln County School District.[7] The only permanent records remaining are the superintendent's biennial reports, which are sparse but somewhat enlightening.

Arda Haenszel moved to Searchlight in 1919 and left in 1922. Her childhood memories are much like everyone's memories of youth, very vivid. She clearly remembers the school and its structure, the wooden boards painted white, the school sitting on top of a hill in the middle of the residential district. The school consisted of two classrooms. During the time Arda attended, there were not enough students to use both of the rooms, so one room functioned as a storage area. The supply of books was limited, and the desks had attached seats, with pencil grooves and holes for inkwells cut into the desktops. These same desks were still being used in the early 1950s. The school had its own very old organ, which frequently needed repair. Even at that time the school was showing signs of the town's decline, for there was only one teacher for all grades. During the three years Haenszel attended school in Searchlight, the largest number of students was thirteen.[8]

Things only got worse for education in Searchlight as the 1920s advanced. The school just barely survived in 1927. According to Nevada law, a town had to have at least three pupils in order to maintain a school. As the 1927 school year was about to begin there were only two children in all of Searchlight. Just in time, however, the John Reid family moved back to Searchlight from Pacific Beach, California, with two school-age sons and a school-age daughter, thus allowing the school to remain open.[9]

The only time since the boom in the first decade of the twentieth century that the school population increased was during the construction of the Davis Dam, from 1946 to 1950. A separate district was formed to take care of the children on the Nevada side of the river downstream where the dam was being built.[10] The growth was short-lived; once the construction ended, the school reverted once again to a single-teacher operation. In all, the school population was large enough to support two teachers in only a handful of years, during the first decade of the century and then sporadically during the decade of the 1930s. The school has been permanently staffed with two teachers only since the 1970s.

In addition to the memories of people like Arda Haenszel, the limited records of the state superintendent give us some information about Searchlight's educational programs. Orvis Ring, superintendent of public instruction, opined in 1902, in language that would fit the tobacco debate of the late 1990s, "Our legislators have passed laws forbidding the selling, giving away or offering to sell cigarettes or cigarette paper to any person under the age of 21 years. Still the habit increases. . . . If there were as much pains taken to protect our boys and girls from pernicious influences as there is to protect the fish and game of our state, a vast amount of good would be accomplished."[11]

Supervision of the many rural schools in Nevada was extremely difficult. In 1904 it was reported that there were 328 schools in Nevada, with 242 separate school districts and an average of twenty-three students per school. The 1904 superintendent's report confirmed the problems of Lincoln County superintendent B. F. Sanders by commenting: "True, there is the State Superintendent, but what can he do towards supervising, when the schools are scattered over so vast a territory. He must attend to the office at the Capitol, look after the deaf, dumb and blind, and perform any other duty the legislature sees fit to impose on him—and that without any increase in salary."[12]

In 1908 it was reported by the Fifth District deputy superintendent, Gilbert Ross, that Searchlight had 101 students and only two teachers;

Mary Lewis was paid $100 per month, and Elizabeth Perkins was compensated at $75 per month. Despite the newspaper reports of fewer than 40 students, the reports indicate that in its early days the school had more than 100.

The high count of students coincided with the height of the mining boom just before the 1910 census. Deputy superintendent John Graham McKay made public in the 1909–10 report to the legislature that Searchlight had three teachers—G. L. Spenser, who was paid $125 per month, Mary Lewis at $100, and Grace Barris at $50.

The 1911–12 legislature found deputy superintendent B. G. Bleasdale reporting: "In schools on the wane, as at Searchlight, in 1911–1912, the school was apportioned money on the basis of 68 students and, during the summer, the leasees' children left, decreasing the school population to 28."[13] Blaesdale also reported that Searchlight was the only school in the district he hadn't visited twice. By 1910 Mary Lewis was the only teacher, and her salary was reduced to $80 per month. This was the first time in ten years that Howard Perkins had not served as school board clerk; those duties transferred to L. W. Goodin. It is important to recognize that people who lived in Searchlight believed there to be far fewer students in school than reported here. The boom/bust nature of mining is one source of such a discrepancy. Another basis for the large number of reported students was the fact that the local school was reimbursed funds by the state on a per student basis. The number reported to the state could vary significantly from the actual number.

Terry Hudgens, a member of a pioneering Searchlight family, recalled attending high school one year in Searchlight. No one could confirm his account, and few believed that he was clearly recalling the facts. But the 1913–14 report of deputy superintendent Bleasdale stated: "High School work is also offered for the first year in Searchlight in Clark County."[14] So Hudgens was right and his critics wrong.

The reports to the legislature did not provide much illumination of the school's woes until 1931–32, when deputy superintendent Ruth Olmstead reported: "The attendance at Searchlight for 1930–31 was 16, and this year the total enrollment was 52. Consequently . . . [*sic*] the mining district has been under severe strain to care adequately for the increase. Possibly for the coming year the enrollment will be less than 20 again." During that period, there was but one teacher, Minnie Peters, who was paid $105 per month.

Searchlight has always had an elementary school. It even offered, for at least one year, high school classes. With the recent growth in the Searchlight area, it seems that another new elementary school will have to be constructed, and even a middle school and high school may be on the horizon.

17

CRIME AND PUNISHMENT, OR THE LACK THEREOF

As a new town, Searchlight had minimal law enforcement. It was far from the well-established main communities of Lincoln County—county seat Pioche, Caliente, and Panaca—all of which were within a few miles of one another.

The only real town in the southern part of the county was Las Vegas, settled as long ago as the Mormon communities of the northern part of the county. Searchlight was the upstart, with little to draw attention away from the better-established areas of this massive county.

The criminal mind thrived in places like Searchlight. Many crimes occurred but were not reported—there was no one to whom they could be reported. The first record indicates that in the winter of 1903 the county agreed to hire two deputy sheriffs at $50 per month each. Deputies W. L. Colton and Light Wheatly began their duties immediately. Within a month the county commissioners decided not to continue paying Wheatly, so the town had to find subscribers to pay the lawman's salary.[1] He was a brave man and on several occasions faced down gun-toting criminals. His police duties, of course, required him to serve warrants, and while he was serving one warrant in 1903, the accused pulled a gun on him. Wheatly said the only way to prevent the arrest would be for the accused to kill him. Fortunately the man yielded to the deputy's bravery.

One of the early historical works on Nevada was written in 1890. The author, Hubert Bancroft, states that from 1846 to 1890 there were 400 murders in Nevada, almost 10 a year. This is a lot of killing, especially when one realizes the sparseness of the population. If anything, Searchlight was less violent than the rest of Nevada.[2]

But killings did occur, the best example being a cowardly murder at Barnwell of George Simmons, the turquoise king and owner of the big mine at Crescent. He was ambushed by William Miller, who in turn was confronted by a lynch mob who threatened to take punishment into their own hands. They were, however, prevented from performing their act of vigilantism.[3] As in the grade B movies that would follow, people of goodwill stopped the mob from executing the accused.

In May 1904, at the opening of a new bar in Searchlight called Shays, a tough guy from the Quartette offered to fight anyone in the house. One man challenged him to a fistfight, another to a wrestling match. Each challenger came forward, and many bets were placed on the outcome. George Baker, the Quartette employee, coldcocked his first challenger and quickly pinned the other. As he was celebrating his victory, he was stabbed in the back by a noncontestant. No lawman was available, and by the time law enforcement arrived the next day, the knife-wielding coward had left town. A search of the countryside proved unsuccessful. (Incidentally, while the fight, wrestling match, and knife attack were in progress, a prospector brought his mule into the bar, and the beast proceeded to eat the buffet that had been intended for the celebrants.)[4]

In Searchlight, as in other western mining communities, personal matters were generally settled with guns. Duels were numerous. In one summer duel over a man's wife, both combatants were killed.[5] When law officers were in town, they arrested people and jailed them, but care of the prisoners was substandard; some even died in jail. One stranger was arrested and then died while incarcerated in 1904; he had no identification when arrested and his name was never known. He was buried in an unmarked grave in the Searchlight cemetery.

Guns became such a problem that the local paper editorialized: "The most pernicious of all habits is that of packing an unnecessary gun or knife. The habit stamps the man as a moral coward: one who does not dare to meet his fellow men on equal terms. He that is known as 'gun man' is held in disrepute and in time of action his reputation works to his own disadvantage; for to be forewarned is to be forearmed. But the worst feature of the habit is someday, acting on the impulse of the moment, the weapon will be used. Then rests the curse of Cain upon him."[6]

Crimes in early Searchlight were varied, and not all involved violence. An eighteen-year-old trammer at the M&M mine was convicted of highgrading and given four months. In highgrading a miner would steal rich ore by secreting it in his pockets, shoes, coat, hat, underwear, lunchbox, or even his hair. Even though only small amounts would be taken from the mine on each occasion, over time it would add up to a nice supplement to the miner's wages. When miners were working high-grade ore, many owners would require them to stop and submit to a search; some even provided showers for the miners coming off shift. In these circumstances a miner wouldn't wear home from work the same clothes he wore to work. It was not uncommon for the miners' bosses to inspect the miners and their possessions when they finished a shift. None of the bosses' deterrents was particularly successful. The ore that this particular young trammer stole assayed at $12,799.55 per ton.

Much has been written about highgrading in Tonopah and Goldfield. It was prevalent in all the mines, done by almost all of the miners. Miners all over Nevada, including Searchlight, developed many ways to steal high-grade ore. Double canvas shirts, made so that ore could be tucked between the two layers, worked well as a device for thievery. Hats with double crowns were devised; they would hold five pounds of ore. Another novel ploy was a long canvas leg pocket that hung down inside the pant leg. Besides these methods, there were always old standbys such as lunch buckets and hollowed-out ax handles.[7] Highgrading caused neverending problems, among them a labor strike in Goldfield in 1907.

The conviction of highgraders in Searchlight was unusual. In Tonopah the court could not convict no matter what the evidence revealed, for many of the jurors were highgraders themselves. One judge even said it would mean his own life if he ever found a suspect guilty.[8]

Mugging is not a modern phenomenon; it often occurred in Searchlight, as evidenced by the mugger who accosted a miner with a gun and shot at him. The miner was not killed, but the bullet caused his clothes to burn.[9] There were also numerous reports of burglaries, blown safes, and embezzling.[10]

One of the most notorious crimes in early Searchlight was the embezzlement of government funds in 1906 by the postmaster, W. B. Atwell, who stole $5,730. He originally admitted taking the money but dared the authorities to prove he had no right to it. To everyone's surprise, he pleaded guilty and was sentenced to four years. The case was the main news in the Searchlight paper for months.[11] The postal inspectors also caught the next postmaster embezzling, but he reacted differently than Atwell had: he committed suicide.[12]

Bert Calkins, a remarkable man by any standard, was also part of law enforcement in early Searchlight, as well as an inventor, assayer, miner, and businessman. While acting as constable during the 1930s, he was notified that a woman was being abused. He hurried to the woman's home and found a man attempting to rape her. Bert called the man by name, and the assault stopped. But the man approached Calkins in a forceful and threatening manner, whereupon the lawman shot and killed him. On that very spot, Calkins laid down his gun and badge, ending his career in law enforcement.[13]

During World War II a man named Seely lived in Searchlight. Known as the town bully, he was employed in supplying Camp Ibis with tires. Like most bullies, Seely always put on a tough front. One night he came into the Texas Cafe with Dot Nellis, the common-law wife of World War II hero Bill Nellis's father, on one arm, and Goldie Epperson on the other. As he passed a small, quiet man named Frank Fairfield, words were exchanged, and as the altercation ended, Seely grabbed his side. After Seely came out of the bathroom he asked Erleen Yeager Givens to look at his wound. When he pulled up his shirt she saw that Fairfield had cut him with his pocketknife, exposing his intestines. Dr. Robert Fenlon, the Searchlight doctor, was called and Seely was rushed to the Henderson Hospital, where he died a few days later. The whole town seemed relieved to be rid of the town bully. Fairfield was never prosecuted.[14]

One notable law enforcement officer who hailed from Searchlight came to town only on rare occasions to visit his family, who were longtime miners in the area; his father was one of the early successful lessees of the Quartette. This man, John Hudgens, became famous as the U.S. marshal of Jerome, Arizona, a town that was lawless until his arrival. It is known that he killed several outlaws in Jerome. He was also a law enforcement officer in Ely.[15]

Another famous lawman was a Clark County deputy sheriff named John Silveria, who was nicknamed Big John because of his very large belly and his expert marksmanship. Silveria was deputy in Searchlight from November 1944 to his retirement in July 1963. He had a simple but direct way of maintaining peace and quiet in the rural Nevada community. Since there was no direct supervision from the sheriff in Las Vegas, he was often both judge and jury. He was the only law of the land, and his only contact with police headquarters in Las Vegas was by car radio. But reception was so bad that he had to drive almost to Railroad Pass, thirty-six miles away, to communicate with Las Vegas.

To make the young think about the dangers of excessive speed, he would load up his squad car with Searchlight youth when there was a fatal accident and take them to see the wreck and the dead body or bodies. He also organized many posses to help him hunt for escaped criminals or others the law was seeking, including people who had gotten lost in the desert.

Searchlight had no legends of notorious bandits, killers, or criminals, except for the infamous Indian outlaw Queho, whose story will be told later. There was never the need for a heavy presence of law enforcement officers; most differences were settled by fistfights, and only on rare occasions did disputes get as far as the courthouse.

18

ACCIDENTS

Mining work is very dangerous, and in the early western mining camps it was much more dangerous than it is now. The state employed a mining inspector, based in Carson City, and assigned deputies throughout the state to assist him. During the first decade of this century, the inspector and his assistants often examined the mining properties in Searchlight. After 1910 their trips to Searchlight, which was some 500 miles from the state capital, became increasingly less frequent. The visits diminished because of the decline of Searchlight mining and the increased activity in Tonopah, Ely, and elsewhere.

Until the leasing era, the Quartette had the finest equipment and supplies available anywhere in the world. There were accidents, but they were few and far between during the days when the big mine was dominant. But after the leasing era began, accidents accelerated at an alarming pace. The Searchlight newspaper rarely reported a mine fatality or mishap before 1909, but many were reported after that. The number of accidents increased because lessees worked very long hours, lacked supervision, were careless, and did not shore up their diggings.

Before leasing, most accidents were typical mining camp incidents. A horse stepped on a piece of dynamite, killing two horses and injuring the driver. A bad piece of fuse caused holes to go off prematurely, causing serious injury. A limb was smashed with a bucket of muck and had to be amputated.[1]

Because the miners in the leasing era were paid not by wages but by the foot or on a percentage, they took shortcuts to get the work done fast. A few examples illustrate the kind of tragedies that occurred when men were desperate for their hard-earned money.

Once, workers were putting ladders down the Duplex Mine, but they were going about the task haphazardly rather than following normal procedure, and a miner named Johnston fell seventy feet to his death.

Another reported accident occurred in the Drake shaft at the Quartette, a development shaft west of the main shaft. A twenty-seven-year-old married man with two children had a habit that led to his death. Instead of using a ladder, he would slide down the hoisting cable—it was quicker but extremely dangerous. Others told him many times to stop the foolhardy practice, but he didn't, and as a result he died.

Another man was killed in a cave-in while doing leasing work at the 1,100-foot level in the Quartette.[2] In November 1910 two accidents were reported. In the one case, a worker was grievously injured by a falling chain. In the other, a miner was nearly killed when a defective fuse caused him to be blasted.[3] In the ensuing years the deaths continued even though work in the mines slowed significantly. For example, as one miner was riding in a skip down the shaft, a cave-in occurred and falling rocks hit him on the head.[4]

In its early days Searchlight, like the rest of the country, had little government supervision of the mines. It was in the large coal mines, where accidents sometimes killed hundreds of miners, that attention began to focus on mine safety. From 1907 to 1909, coal mine explosions in West Virginia, Pennsylvania, and Illinois killed more than 1,000 miners and resulted in the formation of the U.S. Bureau of Mines in 1910.[5]

About that time in Nevada, the dilemma of mine production versus mine safety began to be addressed. Before this time, miners bore the responsibility for their own safety, and it was assumed that death and injury occurred because of the miner's carelessness, inexperience, or negligence.

During the Progressive Era of the 1910s, both in Nevada and in the rest of the country, people looked to government for better protection of mine workers. Searchlight did not benefit from the nationwide movement during its boom years, but just before 1910, the State of Nevada created an inspector of mines post, which was to be a statewide elective office. It was the first step in the state's development of a public policy that would create laws and regulations to protect miners.

When the work picked back up in the early 1930s, the dangers of mining without supervision increased, but the mine inspectors rarely came to Searchlight. Mason Reid and Smoky Pridgeon, both in their twenties, were killed at the Black Mountain Mine in 1935 in a startling mining accident. Pridgeon had just moved to Searchlight from the East and knew nothing about mining. They were working in a tunnel when the holes went off prematurely, killing them both. A teetotaling man got drunk when he was asked to bring the two bodies out of the tunnel of death. He reportedly said, "I had to drink. Those two boys were in pieces."[6]

Bill Hudgens was killed at the Blossom in 1940 when a rock fell on his head. Had he not been working alone, he probably would have survived. But after the rock hit him, he tried to climb out of the mine and fell when climbing a ladder.[7] Many unrecorded accidents resulting in injury and death occurred over the years, until World War II, when mining came to a grinding halt.

Even after the war, mining in Searchlight was without government intervention. I remember working with my father in the mines from the time I was eleven years old through my teens. My father worked alone underground much of the time, which was illegal, of course, but was done for years without a single government citation.

Searchlight also saw many desert deaths and disasters in the early days. Many who came to the desert were not familiar with its perils. In a typical episode, a newcomer, Hal Lewis, tried to walk from a mine in the Eldorado Canyon area to Jean, Nevada, a distance of twenty-five miles as the crow flies. He didn't make it. People didn't understand the vastness and the dangers of the harsh desert environment.[8]

Infections, then called blood poisoning, were also a common cause of death. Depending on the substances in the rock being mined, some of the mines were more hazardous than others. The Blossom had a reputation as a mine where one should avoid any kind of a skin-piercing wound. Even a prick from a desert cactus could cause death if not properly cleansed.[9]

Another common cause of death in the Searchlight area, and probably in all mining camps, was suicide. The loneliness and the austere life of the miner made for a depressing situation. In just a six-month period in 1911, at least two miners committed suicide, one by cutting his throat and the other by shooting himself. One of the men was severely afflicted with miner's consumption; the other had no job and had other money problems as well.[10] Over the years a significant number of miners decided to take their own lives, often because of loneliness, lack of work, overuse of alcohol, or miner's consumption.

19

WEATHER

Searchlight sits at an altitude of 3,540 feet. The town is included in the Mojave Desert region and shares the climatic features of many other parts of the arid Southwest. The sun shines almost 85 percent of each year, or at least 310 days. The average growing season is 277 days.[1]

The vegetation in the immediate Searchlight area is common desert flora, such as Joshua tree, Spanish dagger (or yucca), creosote bush, many varieties of cactus, and scores of other flowering plants. Among the animals are chipmunks, rats, mice, jackrabbits, cottontail rabbits, foxes, coyotes, lizards, wildcats, and some poisonous varmints, including rattlesnakes, scorpions, Gila monsters, and tarantulas.

When one drives into Searchlight from the east, north, or south, it is easy to see that the town is reached only by climbing to an altitude significantly higher than the territory to the east, north, and south. The only entrance to town on a decline is from the mountains of the west, known as Crescent Summit, Timber Mountain, and the McCulloughs.

Accounts in the early Searchlight news report the weather's being surprisingly hot or very cold. In the summer of 1906 it rained continuously from one Saturday through Tuesday of the next week, sometimes pouring and sometimes raining gently.[2]

In 1910 and 1911 there were apparently wide variations in the weather. On a Monday in January 1910, the thermometer hit 18 degrees Fahrenheit, and many thought it could not get colder. The next morning it did—the temperature fell to 10 degrees.[3] The summer of 1910 was just as harsh. In early June the temperature reached a red-hot 114 degrees.[4]

Anyone who has spent time in the desert realizes that wide swings in temperature, as much as 40 or 50 degrees, are typical. Searchlight is no different, as was noted in the early days of the mining camp when the temperature peaked at 114 degrees one day and plummeted to 70 degrees the next morning.[5] Summers are warm in Searchlight, but because of its relatively high elevation, it is cool compared with the surrounding towns and cities such as Boulder City, Cottonwood Cove, Las Vegas, Henderson, Laughlin, Bullhead City, and Needles.

In the summer, temperatures normally reach the mid-90s in the daytime and drop to near 70 degrees at night. An early newspaper reported another high temperature of 114 degrees. But since 1913, when daily records began to be kept, the highest validated temperature has been 111 degrees, in July 1942,[6] even though non-official sources before that time reported at least three 114-degree days.

Winters in Searchlight are pleasant, with daytime highs usually in the mid-50s and nighttime lows in the mid-30s. But when Arctic air reaches Searchlight's moderately high elevation, the weather can turn bitterly cold, as evidenced by temperature readings of 6 degrees in February 1933, 7 degrees in January 1937, and most recently in December of 1990, when the mercury dropped to 8 degrees.[7]

It doesn't rain often, but the area sometimes experiences short, intense periods of very hard rain called cloudbursts (when it appears that the clouds have literally burst). During one of these storms, the rain falls so hard that it is difficult to see more than fifty feet in any direction, and intense thunder and lightning accompany the downpour. Early newspapers made several references to such violent storms. No matter how hardy the soul, these storms are quite eerie. Four or five feet of water can suddenly rush down washes, gullies, and streets. In 1974 when Nelson's Landing in Eldorado Canyon was washed away, survivors spoke of a wall of water more than fifty feet high coming down the canyon. Seven people died in the flood, and automobiles were washed into the river, never to be seen again. The U.S. Park Service was subsequently forced to close the facility.

The annual precipitation in Searchlight is slightly less than eight inches, enough higher than nearby areas to put Searchlight into the semiarid category. The driest months are April, May, and June; the wettest, July and August.[8]

It is unusual for snow to accumulate in the desert, but it does. In January 1949 the ground at Searchlight was covered with fourteen inches of the white stuff, the result of just one storm. That same month an incredible thirty-eight inches accumulated. Any adult who was living in Searchlight at the time vividly remembers the hardships imposed by the violent and rare winter storms. Among those killed on the highways was one of Searchlight's leading citizens, Babe Collins. The mail was not delivered for days. Cattle died by the hundreds. Roads were impassable and people had to walk through the snow to get to work.[9] This same storm wreaked havoc on the livestock industry as far north in Nevada as Elko County.

Records of the weather in Searchlight have been kept steadily and consistently since 1913, when postmaster C. L. Mohn began the process. He continued until 1923, then Ann Crowley took over for the next three years. Light Wheatly had many different duties in Searchlight from its very inception, and one of his functions in his later years until his death in 1931 was keeping the weather records.

Bernice Kamer, who would later marry Dr. Fenlon, kept the weather information from 1931 until 1941, when the Nevada State Department of Highways took over the task and built a large maintenance yard in

Searchlight to take care of the newly paved road from Las Vegas to the state line, thirteen miles below Searchlight. The Highway Department has been the official keeper of the records now for more than fifty years.

A look at the lows and highs for each month of the year will give a sense of the variations in Searchlight's weather. The highest temperature for any January occurred in 1971 when the thermometer hit 71; the lowest January temperature was in 1937 when it dropped to 7. In February the high was in 1986 at 81, the low in 1933 at 6. March's high was 90 degrees in 1981, the low was 20 in 1966. For April, 1982 had the record high, with 94 degrees, and 1945 had the April low, 27 degrees. May has had many high marks, but the record was made in 1974 when the temperature hit 104 degrees; an all-time low for that month occurred in 1915 at 30 degrees. June's record high was established in 1924 when the temperature hit 110; a low record of 40 was set in 1932. July hit 111 in 1942; the low for that month was 52 in 1987. August warmth held with the high reaching 110 in 1933, and in 1920 the low was 51 degrees. September hit 107 in 1950 and slid to 41 in 1986. The hottest day for an October reached 98 in 1978; the coldest was 23, in 1971. November was still warm in 1986 when the temperature hit 86, but the record low was way down to 15 in 1919. December hit a high of 78 in 1928 and a record low in 1990 of only 8 degrees.[10]

Lake Mohave, though only fourteen miles away, is at least 10 to 15 degrees warmer than Searchlight itself, and the Timber Mountains, only ten miles to the west, are usually 10 to 15 degrees cooler. Because of its clear air and moderate temperatures, Searchlight has in recent years become a haven for retirees.

20

ACADEMY AWARD CHAMPION

An insight into the transition between Searchlight's boom and the beginning of its economic slide comes from the woman nominated for more Academy Awards than any other person, costume designer Edith Head. Edith's stepfather, Frank Spare, first came to Searchlight around the turn of the century from Mexico, where he also was engaged in the mining business. Edith spoke highly of her stepfather and seemingly had an excellent relationship with him. He was hired as the superintendent and general manager of the Pompeii, a mine located just north of the Blossom.[1] His name appears on the voter registration rolls for 1904.

The Pompeii was never a big producer, but it had been the center of various levels of activity for many years. The mine closed in 1904 but reopened again, as it would do periodically for the next twenty years. As one of the few mines in Searchlight that had a vertical shaft, the Pompeii was quite distinctive. The shaft was dug straight down to a depth of almost 700 feet.

The company built a beautiful home for the Spare family at the Pompeii in 1905. The pretty bungalow was painted moss green with white trim; the home had five rooms and deep verandas all around. It was described as the prettiest house in the district.[2]

In her autobiographical writings, Head confirms that Searchlight was named as a result of the offhand remark that Colton needed a searchlight to find the gold in his ore.[3] Head remembers the bungalow differently than when it was first constructed, however. One imagines that it could have deteriorated significantly in the desert sun from the time it was built until she left around 1910, or she may have understated the description simply to dramatize her austere beginnings for her fans and Hollywood friends. She described the Searchlight house as "unpainted wood with a porch on three sides, with a slanted tar paper roof, with a jutting stove-pipe."[4] More likely than not, her narrative was exaggerated for dramatic effect. Her own biographer found her to be secretive, often telling different versions of the same events. For example, she tried to keep her maiden name, Posener, a secret. She claimed she never went to grammar school, but that is simply not true; she graduated from grammar school in Redding, California. In one of her books she states that there was no school of any kind in Searchlight.[5] In short, this great woman, for whatever reason, exaggerated and on some occasions altered the facts to create the image that she felt she should have in Hollywood.[6]

Edith Head was one of a number of significant Searchlight residents who would become famous. She was the first woman to become a noted clothes designer for Hollywood's elite, and she was a leader in Hollywood film fashion for six decades.[7] As a noted costume designer, she was nominated thirty-five times for the Academy Award, and she won six straight. She designed wardrobes for almost every Hollywood star, including Audrey Hepburn, Elizabeth Taylor, Grace Kelly, Gloria Swanson, Bette Davis, Hedy Lamarr, Angela Lansbury, Olivia de Havilland, Barbara Stanwyck, Loretta Young, Dorothy Lamour, and Robert Redford.

Head clearly remembered various Searchlight experiences such as going down in the Pompeii and watching the men at different levels eat lunch. For entertainment her stepfather would take his only child down the mine on the elevator (as she described it).[8]

Edith Head came to town after her stepfather arrived, in October 1903, and celebrated her sixth birthday in Searchlight on October 28. She arrived when the town was at its economic pinnacle. She spoke well of Searchlight in several interviews, in spite of her statements on other occasions that made her home there seem more austere than it actually was. It seemed that she wanted some of those she had business with to believe that her young days were poverty-stricken, but the facts were quite the opposite.

Mitch Fox, a Las Vegas journalist, interviewed Edith Head several years before her death in 1981 at the age of eighty-five. She remembered Searchlight as being rowdy and intemperate. She expressed her utter delight that the old camp was still in existence, as she had not returned since she left as a little girl.[9] She was obviously proud of her past.[10]

During the early 1970s, I happened to sit next to Edith Head on a flight to Reno. We had a wonderful visit—two people, for an hour, discussing memories of their youthful days in the mining camp of Searchlight. That conversation is one of my most pleasant memories.

Shortly before her death on October 20, 1973, in response to an invitation to a Searchlight celebration, she wired Gayle Colton as follows: "only the fact that I have to go to Italy keeps me from joining you on the 25th. As one of the earliest residents of Searchlight, please let me extend my congratulations and best wishes for a wonderful occasion, and please tell everybody I still feel very proud of my early days in Searchlight. I am sending you a small donation and hope that in the near future I will be able to visit my hometown."

Edith Head also had a more indirect connection with Searchlight. The first big star that she dressed, in 1927, was the "It Girl," Clara Bow, who would later retire to Searchlight with her husband, Rex Bell.[11]

21

HOW BIG WAS IT?

F. W. Oakey came to Searchlight in 1901, when he was nineteen. He left after about three years, at the height of the famous Searchlight strike. He recalls the buildings in early Searchlight, when even the jail was a canvas tent. He remembers the strike that almost permanently closed the mining camp and could describe the buildings that were there as of 1903—constructed with wooden frames and corrugated iron on the sides.

The Searchlight that Oakey depicts was before the coming of electricity. Kerosene lamps and lanterns gave light. There were few coal oil lamps, and most heating and cooking energy was provided by wood or coal. The coal had to be brought in by teams of animals from the railroad at Barnwell, and the only wood available was catclaw or mesquite.

This wood could be gathered only in the washes, and the only parts of the mesquite that could be used as fuel were the dead branches. Cedar and pine wood, always sparse, came from Timber Mountain, ten miles to the west.

Oakey said that when he left the new town, the population was about 600.[1] His figure is similar to that given in the summer of 1905 by the local newspaper.[2] Curiously, though, just a week previously, the same paper stated the population to be 1,200. And the next summer the paper opined that the number of people living in Searchlight was 1,500.[3] Earlier in the year when the Santa Fe was trying to decide whether it was going to build a railroad into Searchlight, a letter between company officials indicated that the population was 1,000. The individual who prepared the letter based his statement on hearsay, as he didn't live in the town and had probably never been there.[4]

Bob Kerwin, who lived in Searchlight from 1908 through 1915, wrote of his experiences there in the *Nevada Historical Society Quarterly.* By the time he arrived, the town was already on a downhill skid. The two-room school had only one teacher, and by the summer of 1909 the *Bulletin* estimated that the population had fallen to 1,000.[5]

In 1906, however, the voter registration in Searchlight was almost double that of Las Vegas and also Caliente.[6] In the previous general election Searchlight had polled more votes than any other place in Lincoln County.[7]

We can determine with some accuracy the number of people living in Searchlight in 1900, 1910, and 1920, since the U.S. census data are now available for review. No more recent census information is available because federal law prohibits publication of the information until seventy years after the date of the previous census. The U.S. Manuscript Census of 1900 states that there were 169 non-Indians in Searchlight and 42 Piutes. In 1910 the accounting lists only 13 Indians and 600 others.[8] Other sources based on census reports list the 1910 population as 387, making Searchlight the smallest city in Nevada.[9] The 1920 census put Searchlight just below the 200 mark. Phillip Earl, a renowned Nevada historian, lists the 1930 population at 137.[10] Through the census information supplied, we can reach several conclusions.

In 1900 a large number of occupations was listed, but only 17 wives, 2 sons, and 9 daughters. Of course, the single most frequently listed occupation was that of miner. Eight grocers, 8 teamsters, 7 carpenters, and 22 laborers appeared in the census.

The 1910 census indicates that 108 individuals, or about 20 percent of Searchlight residents, were boarders. But even with this available information it is still difficult to come up with a definitive number for the town's highest population.

The first census taken in Searchlight, in 1900, showed there to be 30 foreign-born residents; these 30 plus the 42 Indians made a significant minority. Ten years later, after Searchlight had reached its peak in population, of the 600 residents within the town, 90 were foreign-born; the 1920 census listed only 16 as foreign-born. Many parts of the world were represented in these three census years, including France, Germany, Switzerland, England, Wales, Austria, Slovenia, Scotland, Lithuania, Sweden, Denmark, Norway, Ireland, Canada, Mexico, Japan, and China.

The town had no ethnic sections, nor did the graveyard. In some Nevada towns, such as Ely, there were ethnic neighborhoods and even segregated burials in the cemeteries.

The numbers cited by all of the sources discussed above do not include the populations of Camp Thurman, the river settlements, Crescent, or any of the smaller mining camps such as those at the Buttes, Nob Hill, or Red Well (formerly called Summit Springs). These locations had significant numbers of people over the years, with fluctuations depending on the levels of activity.

It seems clear that Searchlight's population reached its peak in late 1907 or very early 1908. Phillip Earl lists the 1907 population at 6,000.[11] There seems to be no basis for this large number, even though mining booms resulted in hundreds of flash-in-the-pan towns, with populations reaching into the thousands before the towns rapidly withered away when the resources did not meet expectations. The boom-and-bust cycle could pass very quickly. Weepah, Nevada, for example, grew from nothing to a town of 1,200 in 1927, even though the claims were actually worked for only about two months.[12]

So any affirmation about the largest number of people to live in the Searchlight area at any given time is merely an educated guess. A more accurate calculation is not possible because the height of Searchlight's boom occurred at least two years before the 1910 census. The "camp without a failure" began to fail at approximately the same time the railroad arrived. During the last few weeks of the railroad's construction, 300 people lived in or near Searchlight, at Crescent, Camp Thurman, and the other outlying homesteads and small mining camps. The railroad also significantly increased the population around Searchlight for a few months because of construction crews. The Searchlight area, including the camps at Crescent, Thurman, Red Well, and the Buttes, attained for a few weeks a population of about 3,000 people, excluding those who lived in Eldorado Canyon. It would be difficult to establish any population of a larger number at any time.

The CAMP WITHOUT A FAILURE did not fade entirely into oblivion, as did many Nevada mining camps. It survived partly because small amounts of ore remained for the lessees and other men who toiled in the earth for little return. But the main reason for the survival of this unique town was that its economy did not have to depend totally on ore deposits. The town became the center of a main highway arterial, located as it was on the Arrowhead Highway until 1925, when what is now Interstate 15 was bladed between Barstow and Las Vegas. The tourist trade helped keep the business community alive, and prostitution also helped to make the town a destination resort. Later, the commercial development of Lake Mead National Recreational Area and Lake Mohave, after the construction of Davis Dam, and the development of the town of Laughlin would also be commercial boons.

Today the population of Searchlight, including Cottonwood Cove and Cal-Nev-Ari, stands at more than 1,000. In the last election, almost 500 residents actually voted. The town is slowly but surely growing and in the next couple of decades will match the population of 1907.

22

FLYING HIGH AND FLYING LOW

Shortly after Edith Head left Searchlight, the town reached such a low ebb that by 1912 the once-promising mining community could no longer even pretend to be prosperous.[1]

It is difficult today to comprehend that the thriving metropolis of Searchlight even had a song written in its honor by one of the most famous composers of all time, Scott Joplin. In 1907 Joplin wrote and published "Searchlight Rag," named after the Nevada mining town. The creator of such well-known popular pieces as "Maple Leaf Rag," "The Entertainer," and "Fig Leaf Rag" actually never set foot in Searchlight. Even though it cannot be proved that he ever came west of the Mississippi, there are several who say he played the piano in Searchlight. This, they say, is why he wrote "Searchlight Rag."

A careful reading of the several detailed accounts of Joplin's life, however, never gets him farther west than the Mississippi.[2] But one of Searchlight's real characters, Rags M'Goo, claims that Scott Joplin played the piano in Searchlight on July 4, 1914. According to M'Goo, "The mining boom had just started to wind down, but there still was 3,000 people in town. I went up and stood along side of him, just to watch him play."[3] There is no question that M'Goo was a young boy in Searchlight at the time, having arrived there either in 1911 or 1913.[4] It is obvious also that M'Goo learned to play the piano in Searchlight and played it for a living his whole life long. His professional career took him all over the United States to play ragtime. He left Searchlight in the early 1920s and didn't return except for brief visits until he was nearly eighty years old. All credible evidence indicates that his statements about Joplin are as much without foundation as his estimate of the population of Searchlight in 1914. There is, however, sound evidence to support Rags's assertion that the piano he learned to play on in 1914 Searchlight is the Kohler upright that is still in use in the community center.[5]

The weight of evidence shows that "Searchlight Rag" was written by Joplin as a favor to a nightclub owner and one of his employers, Tom Turpin. Turpin, like Joplin, likely never visited Searchlight. But he did have a mine called the Big Onion, near Searchlight, in the Camp Thurman area. The *Bulletin* gave some credence to this ownership when it obliquely observed, "The Big Onion, owned by a colored man, shows high values in gold and silver, as high as $140. $30,000 is said to be offered for this property."[6] Some proffered that Turpin had bought some stock in the mine called the Big Onion, which produced nothing other than some money for the promoters of the mine when they sold their stock to investors like Turpin.[7]

During the years 1910 to 1917 little positive information about Searchlight is to be discovered—except the cheerleading of the *Bulletin,* and even this ended when the paper went broke in 1913. The town lost its economic vitality during these years of steep decline following the initiation of leasing by the Quartette in 1908.

In a last-ditch effort to showcase Searchlight, the stalwarts of the town produced the *Searchlight Review,* a single-issue tabloid newspaper written by George Corn and published under the auspices of the Searchlight Chamber of Commerce in 1917. This one-time magazine-like publication gave a retrospective review of Searchlight, which was twenty years old at the time. Since there was no newspaper in town then, the *Searchlight Review* provides an important perspective on Searchlight that we would otherwise not have.

When the *Searchlight Review* was published, the Chamber of Commerce was fifty-four members strong and included many of the real pioneers from the early days of the camp. Among them were Light Wheatly, a man who had been a jack-of-all-trades in the mining camp, a law enforcement officer, a mining contractor, and a hotel operator; H. A. Walbrecht, who had been the utility king, having owned or operated all of the utilities in town at one time or another; James Cashman, who is prominently featured throughout the publication; A. C. "Bert" Calkins, an assayer and inventor who, when married, honeymooned in the private car of Collis Huntington, builder of the Los Angeles–to–Salt Lake portion of the Union Pacific Railroad.[8]

Also listed as members of the 1917 Chamber of Commerce were some newcomers to Searchlight, among them Sid Gaines and Chris Kirkeby. These two men became responsible for most of the employment in Searchlight mining during the 1920s and 1930s, by virtue of their leasing operations on various properties, including the Quartette and the Blossom.

One of the new Searchlighters, who is mentioned as a member of the Chamber of Commerce, earned the distinction of full-page coverage, including a picture: James John Macready, son of Benjamin, one of the original pioneers of the settlement. John is listed as still being in his twenties, a graduate of Stanford University, and a twice-elected justice of the peace.

Among many of the monied class in these frontier towns, the husband would live in the mining camp, but the rest of the family would live in a larger city such as Los Angeles or San Francisco, where schools and other services were more in keeping with what they felt to be their rightful place in society. Benjamin Macready followed this practice with his own family. Thus John Macready (only in the Searchlight 1917 publication is he referred to as James) was born in San Diego at some time before his father began the Quartette in 1897.

Often the personal section of the Searchlight paper carried announcements about the comings and goings of the bosses and their children and other family members to and from the mining camp. In 1912, after his graduation from Stanford University, John Macready arrived to spend the summer with his parents at the Santa Fe.[9] This trip was actually intended to be a brief period of recovery from the rigors of school, after which he was to continue on his way to the East Coast to be employed as a stockbroker.[10] While this handsome young man was relaxing in Searchlight, the townspeople convinced him that he should run for office. He did so and was elected twice, in 1914 and 1916, as the justice of peace in Searchlight.

Before this trip, he had come to Searchlight for the summer and for other vacations on many different occasions and had often worked in his father's mine. Benjamin always wanted John to work like anyone else and not seek soft jobs just because his father was the boss. Young Macready was therefore respected as someone who did not try to be better than others simply because he had a wealthy and distinguished father.

After relinquishing a career as an East Coast businessman, John Macready did a fine job in his position as part-time judge. In Nevada, a justice of the peace handled preliminary hearings on felony cases and presided over misdemeanor cases, as well as having the power to perform marriages. John Macready also went into business in Searchlight, trying his hand at freight, ferrying, and mining. While he was justice of the peace, one of the town tough guys challenged him to an off-duty fight. Macready accepted, and they set up a ring for the match. The judge knocked out his opponent and won the fight, but to everyone's surprise, his challenger had a glass eye, which popped out as Macready landed the final blow![11]

Benjamin Macready had made his mark as a founding father of Searchlight and as an adventurous western entrepreneur. Before 1920 he moved back to California and went into the newspaper business. His Southern California newspaper was later purchased by the *Los Angeles Times*.[12]

But Benjamin Macready's son would match and many times exceed the father's pioneering spirit.

John Macready soon became a pioneer in his own right. With the outbreak of hostilities leading to World War I, young Macready traveled to Reno in 1917 to join the U.S. Army Air Service. He served with little distinction during the war, but he learned to fly, and after the war he flew as no one had flown before and very few have since.

The most prestigious award in aviation is the Mackay Trophy, given each year by the National Aeronautique Association. John Macready, the former justice of the peace of Searchlight, won this coveted award three times—a feat accomplished by no one else in aviation history. This hero of American aviation first garnered the award for setting an altitude record of 34,509 feet, a remarkable achievement, considering that he was flying in an open cockpit. For the next six years, he set altitude records as the chief test pilot for the U.S. Army Air Service, even going higher than 40,000 feet, where temperatures sometimes dropped to near minus 100 degrees Fahrenheit. During these pioneering high-altitude flights, Macready breathed oxygen through his mouth by means of a welder's mask. The keeper of his records for these historic events was Orville Wright.[13]

Macready won his second Mackay Trophy for setting an endurance record by piloting an airplane for more than thirty-five consecutive hours. This historic flight also featured the first-ever inflight refueling. The famous adventure occurred on October 2, 1922. The record was made possible as a result of an aborted attempt to fly cross-country, which failed because Macready and the copilot could not reach the proper altitude after leaving from their San Diego base.

In 1923 Macready flew a nonstop transcontinental flight across the United States and thereby secured his third Mackay Trophy. He and his copilot had attempted the cross-country flight twice before, taking off from San Diego and flying east. On the third and finally successful flight, they took off from New York and flew west. It took Macready and another pilot, Oakley Kelly, to do such daredevil flying. The plane was a foreign-made craft that had been retrofitted. It was redesigned to carry almost 800 gallons of fuel, more than twice what the plane was originally designed to carry. The engine was classed at 400 horsepower at sea level.

The lead pilot sat high up in the nose of the plane, with the engine by his side and partially under his right arm, an arrangement that allowed him to make minor repairs during flight. Another control was positioned ten feet to the rear, but it was nearly impossible to fly from behind, as that pilot's view was blocked except for a limited line of sight from the door to his left.

The two stations each had room for only one pilot. The pilots communicated by written message, delivered by one of them crawling through a tunnel used for the communication wiring. Another way to communicate was for one of the pilots to jerk the plane's controls, thus notifying the other that it was time to exchange positions. They actually changed places five times during the flight to San Diego, by pulling out the back of the pilot's seat and crawling through the narrow passageway. This pioneering transcontinental flight had enough thrills to satisfy most pilots, but not Macready.[14]

A 1924 *National Geographic* included a fine full-page aerial photograph of Searchlight, Macready's former home, taken by him during the extraordinary flight.[15] The importance of his transcontinental flight cannot be overstated. As one aviation expert stated: "The historic flight of the U.S. Army Air Service's T-2 is one of the greatest flights in the history of aviation. The flight is visible evidence of the state of art in 1923 in which men set out on an untried trail, pitted themselves and their craft against the elements, and won against all odds."[16] Lowell Thomas described Macready's effort as one of the famous first flights that changed history.[17] Within a few minutes after landing in San Diego, the pilot received a telegram from President Warren Harding reading, "You have written a new chapter in the triumph of American aviation."[18]

The airplane flown from New York to San Diego by the Searchlight justice of the peace is now prominently displayed at the Smithsonian Institution's Air and Space Museum in Washington, D.C. It looks surprisingly large and powerful, from its perch overlooking Lindbergh's *Spirit of St. Louis* and other Smithsonian treasures.

An early aerial photographer, John helped to popularize the Grand Canyon with some of his 1923 photographs of the great natural wonder.[19] He was also the first pilot to parachute from an airplane at night—not by choice but because his plane caught fire.[20]

In 1921 John Macready had become the first to use an airplane for crop dusting. This Pacific Coast lightweight boxing champion was also awarded the Distinguished Flying Cross and the French Croix de Guerre with palm, he became a member of the Society of Experimental Test Pilots, and he was elected to the Aviation Hall of Fame. Not only did he see service in World War I but he was called back into the Air Corps during World War II.[21] Throughout his magnificent career, John Macready never forgot Searchlight and returned on occasion.

I remember describing John Macready's amazing feats to war hero, astronaut, and U.S. senator John Glenn. Senator Glenn responded that once when he was at a gathering of aviation buffs at the Aviation Hall of Fame, he had talked to Macready. By this time Colonel Macready was an old man, but he was still very alert. Glenn asked Macready why he had flown from the East Coast to the West? Did he not know about the prevailing winds that always blew from west to east? Macready, in a huff, responded quickly, "We tried twice and damn near killed ourselves both times. We couldn't get over the mountains." Glenn felt that Macready believed everyone should have known the torment he went through to be the first to fly across the continent.

The 1917 Chamber of Commerce article did its best to throw a positive light on Searchlight. This puff piece is never more overstated than in its description of early morning in the town: "We see the green fields and the flocks and herds as they rise to the morning meal and as we stop for one last look." Anyone who knows Searchlight would realize that such a statement is a real exaggeration. There is not now, nor has there ever been, a green field in Searchlight.

In several places the publication attempts to make Searchlight unique. For example, it states that the "wild cat" period in the history of Searchlight had passed, meaning that prospectors were not discovering new mines. The document also casually states: "Searchlight is not a boom town today." This may have been hard for the people of Searchlight to admit after some years of prosperity and almost ten years of attempting to make the town appear to be more successful than it really was.

Trying to make the situation look brighter, the piece also indicates that the war situation in Europe had improved mining activity all over the United States, especially in Searchlight. The mine section of the report informed readers that 1916 was a better production year than 1915 and that there was a 50 percent increase in the number of Searchlight claims.

The report noted that the school had nine grades, a statement that is supported by other evidence. As previously noted, one longtime Searchlight resident stated that he attended a year of high school in Searchlight. Before the discovery of this article, there was no independent corroboration that his claim was correct.

We do know that Searchlight's fortune did not greatly improve during the next two years. Arda Haenszel lived in the town from 1919 to 1922 and describes these three years in graphic and sentimental detail in her 1988 book.[22] She was a valuable source of information about the Searchlight of the 1920s. In 1996 she was asked to review the 1917 Chamber of Commerce publication, in order to find out if it would jog her memory regarding other details. She said that what struck her the most was the changes that had taken place in just two years, i.e., from the publication in 1917 of the Chamber of Commerce piece to her arrival in 1919.[23] All her comments focused on the dramatic changes in those two years.

Haenszel also remembered persons mentioned in the 1917 publication—for example, Willet Barton, whose daughter, Wilberta, gave Arda violin lessons. Barton, a graduate mining engineer, later became a mining operator and a large landowner. His notoriety, however, came not from his business prowess but rather from his murder in 1939 of Bill Stiles, an alleged claim jumper. Though Barton was convicted, he served only one year in jail for shooting down a defenseless man working in a hole. His story was that he shot into the ground and the shot ricocheted up and hit Stiles.

Arda also remembered Ollie Thompson, assayer and protector of the

Cyrus Noble Mine. Ollie lived on or near the Cyrus Noble until he died in 1960.

While Arda lived in Searchlight, there was no church or church services, though various townspeople would occasionally teach Sunday school for the children. Arda remembers swimming in the M&M's water tanks. She also recalls that no elected officials visited the town after its decline. From 1919 to 1922 there were two hotels, the Nevada and Wheatly House. The railroad ran only two days each week, and there was no ice, laundry, meat market, barbershop, or homegrown meat or produce.[24] The only mercantile left in town was the Searchlight Mercantile, owned by Ella Knowles.

Arda's father, a doctor, came to Searchlight seeking a dry climate for treatment of his tuberculosis. Dr. Haenszel was the only physician between Las Vegas and Needles. His predecessor, Dr. Hastings, had been in Searchlight for many years, but he left when the community went into decline. Dr. Haenszel initially kept fairly busy with the railroad personnel twenty-three miles away at Barnwell, with the few ranchers in the area, the three shifts working in the Techatticup Mine at Eldorado Canyon, and the few cases coming from Searchlight itself. By this time there was no hospital or nursing staff in the whole region.[25]

In the preface to Arda Haenszel's book, Dennis Casebier, historian of the California, Arizona, and Nevada deserts, states: "Searchlight wasn't a ghost town, but was 'ghosting.'" One of the factors contributing to the town's downturn was prohibition, which went into effect on January 16, 1920. By then the downtown area was severely depressed. Many of its buildings housed saloons, and when the Eighteenth Amendment to the U.S. Constitution was passed and those establishments closed, much of downtown Searchlight went dark.

It is not known how many in Searchlight cooked moonshine or illegal whiskey, but it is clear that many Searchlight residents had stills at several locations on the river and elsewhere wherever there was water.[26]

Arda Haenszel liked the railroad station because it was relatively large, with upstairs living quarters for the agent and his family. She remembers that the few remaining commercial establishments in Searchlight were those on the main street. The largest structure was the Nevada

Hotel, formerly Kennedy House. On the first floor was a soda fountain for all, and liquor was always available for any adults who were interested. The Searchlight Hotel had burned by the time Arda came to town. Except for gasoline and car accessories, the Searchlight Mercantile, owned and operated by Ella Knowles, sold everything she was able to stock during these times of economic hardship. There were no specialty shops such as drugstores, dry goods stores, hardware stores, or bakeries. There was no cafe, but one could get board at the Nevada Hotel or Wheatly House.[27] Houses were lighted by kerosene lamps or by candles. It was obvious to Arda that at one time the town had had electricity, for there were power poles and indoor wiring in most of the buildings. During her time in Searchlight, her family's electricity was supplied only during limited hours by a private gasoline generator that also supplied power to some of the businesses.[28]

One thing that Dr. Haenszel noted was the lack of fresh milk. The local store carried all canned goods, beans and other nonperishable items, and it also carried various mining supplies like nails, shovels, and the like. The only way to preserve food was to use a desert cooler—a box covered with burlap that was continually kept wet; the evaporation and the desert breeze cooled the contents of the box. This method of cooling in Searchlight extended into the early 1950s.

Haenszel writes of an outing to Piute Springs, a respite that people in Searchlight have enjoyed since it was discovered. No narrative about Searchlight would be complete without mention of this oasis in the desert, about fifteen miles south of the mining camp. Piute Springs was inhabited long before gold was discovered in Searchlight. The springs are in a heavily vulcanized small mountain range that rises to about 2,500 feet above the Colorado River, which lies twenty-two miles directly east of the springs. Just over the mountain range to the west is Lanfair Valley.

Those who spent time in Searchlight, like Arda Haenszel, always recalled Piute Springs with fondness. In early Searchlight it was the only water within twenty-five miles.[29] The trip to Piute today leads one about ten miles down the road, to Needles, California, almost to the state line, then a couple of miles west, then south again on the power line road for a few more miles, with a short jog to the west. As the crow flies, the springs are almost twenty miles from Searchlight.

Almost miraculously, this stark rocky formation spews forth about 250,000 magical gallons of pure fresh spring water every day.[30] The water flows west for more than a mile toward the mighty Colorado, then sinks into the thirsty desert in a myriad of plant life, including cattails, grasses of various species, mint, water lilies, and desert flowers.

The bird population at the springs, described by a visitor in the 1940s, was vivid and incredible: the sound of frightened birds leaving the wash of water was like a jet plane taking off. Among these hundreds and hundreds of birds were quail, dove, vultures, chickadees, sparrows, jays, warblers, hawks, falcons, robins, and many other species.

The first humans to use this oasis in the desert were Native Americans; their encampments and forays are still evident despite years of thoughtless vandalism. Indian writings are still present, but they have been significantly diminished. The tribes who ventured into this garden spot were Piutes and Chemehuevis. They even grew squash, corn, and melons. Traces of ancient trails in the area reveal the likelihood that Piute Springs was a popular gathering spot.

The first non-Indian to visit the site was the famous Spanish explorer and Catholic priest, Father Francisco Garcés. He came in the same year as our American Revolution, 1776. Even though he did not specifically mention the springs, it is clear from his writings that he was at or near the area on May 8, 1776.[31]

The first written description of this land came from a railroad exploration party led by Lieutenant Amiel Whipple in 1854. Whipple, in fact, was one of three in the party who left written accounts of this exploratory trip, and he was the one who named the springs. He also reported to his superiors that because of the site, either a railroad or a wagon road could be developed.[32] Whipple became a major general in the Union Army and was killed at the battle of Chancellorsville in 1864.[33]

Three years after Whipple's trek, the U.S. Congress appropriated monies to develop the already established Mojave Trail. This venture was led by Edward Beale, a retired naval officer. Beale was given the task of evaluating twenty-five camels in terms of their ability to adapt to this part of America. It took four years to make the route permanent, but it was rarely used.[34] The U.S. Postal Service used the trail for only two years as a mail route. Beale's route was in service for four years and led to the permanent Mojave Road. Because the Mojave Indians continually harassed travelers on this route, Fort Mojave was built at the point at which the road crossed the Colorado River. The fort was abandoned during the Civil War.

The trail was used most heavily in 1863–65, because of the discovery of gold and silver in Arizona and, of course, because of the Civil War. Even though thousands of soldiers and others passed over the Mojave Trail during this time, very few spent more than a day or two at the springs.

Most of the commentary about the springs during these early years revolved around the difficulty of getting over the pass, which had a reputation of being one of the worst hills on the wagon road west.[35]

In November 1867 construction of a U.S. Infantry station was started at the springs, intended to protect the mails of the United States. The fort consisted of five rooms plus a corral. Construction costs for the moderate-sized facility were $477.50. The work on the fort took from November 1867 to May 1868, and fewer than twenty soldiers handled all of the construction. Soldiers stationed here provided escorts for riders on the mail route from San Bernardino to northern Arizona.[36]

The original construction plan, including site location, was coordinated by two military officers who came to the area on assignment from the U.S. Army. Major Henry Robert would be remembered not for his work on the fort but for his work on parliamentary procedure: he was the author of the famous *Robert's Rules of Order.* Samuel Bishop, a civilian trader and general businessman, would later have a city named in his honor: Bishop, California. In 1974, more than a hundred years after Bishop left the fort, a large rock with his name carved on it was discovered in a wash just below the site.[37] This boulder is now in the San Bernardino, California, museum.

The fort was abandoned shortly after it was built, and we have little knowledge of the doings at the springs until gold was discovered in Searchlight. It was not until the early 1920s that 140 acres were homesteaded and part of the site became privately owned. But by World War II the site had been abandoned, and the California Department of Fish and Game now owns the formerly private land.

The fort was in excellent condition until the 1920s when vandals and souvenir hunters slowly ruined the structures as they were originally developed. Even in the 1950s the walls of the fort and many of the rock paths were in good shape. Today, however, little remains. What is left, however, is the never-ending mystery of the pure spring water that visitors like Arda Haenszel always remembered. Those travelers referred to Piute Springs as the oasis in the desert, and people who lived in Searchlight before 1960 still speak of Piute Springs with reverence.

During the three years that the Haenszel family lived in Searchlight, they saw the progressive deterioration of the town, economically and otherwise, as the mines shut down one by one and people continually moved away. Dr. Haenszel and his family went to San Bernardino, but Arda Haenszel, a retired California schoolteacher, gave us a view of Searchlight that would otherwise have been lost forever.

Big John Silveria at his home in Searchlight. Courtesy *Las Vegas Review-Journal.*

Bill Nellis after high school graduation, 1936. Courtesy Gary Nellis.

Colonel John Macready with his experimental airplane, ca. 1925. Copyright © 1990 Macready Foundation. Courtesy Macready Foundation.

John Macready (center) with his parents, Mattie and Benjamin Macready. Copyright © 1990 Macready Foundation. Courtesy Macready Foundation.

Searchlight jail, ca. 1959. Courtesy Joyce Dickens Walker.

FIRST NATIONAL BANK

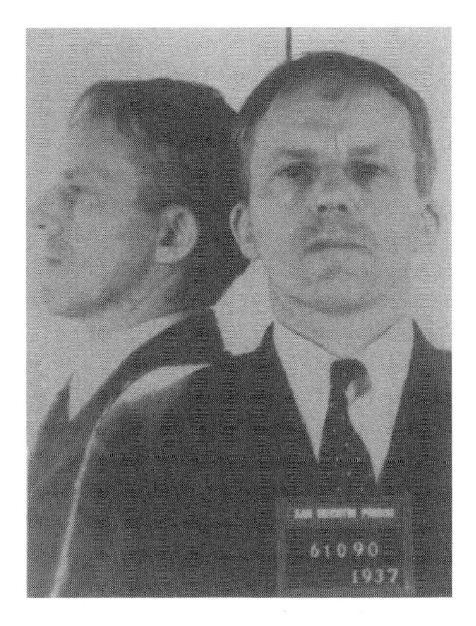

Cancelled check from Homer C. Mills to Pete Domitrovich, September 30, 1951. Author collection.
Prison photograph of Homer C. Mills, 1937. Courtesy California Department of Corrections.

After thirty years, Queho is finally captured. Charlie Kenyon (left) holding Queho's double-barreled shotgun, Sheriff Frank Waite (center) holding Queho's skeleton, and a park ranger believed to be Guy Edwards (right) holding Queho's .30-.30 rifle. Courtesy *Las Vegas Review-Journal.*

Queho's remains as found in his Colorado River cave, 1940. Note left leg wrapped in burlap. Loaded and cocked .30-.30 stands in entrance from which anyone approaching could be seen. Ledge in back of cave is covered with loot taken from miners and other victims. Paddle-like implement was used to slap water in imitation of a beaver's actions; area was full of beavers and sidewinders. Courtesy *Las Vegas Review-Journal.*

A dramatic fire in downtown Searchlight, 1949. Courtesy Joyce Dickens Walker.

Willie Martello, king of prostitution in Searchlight. Courtesy Marshall Sawyer Jr.

Buddy Martello (Willie's brother) with Christine Jorgensen, the first male to undergo a transsexual operation, at the El Rey, ca. 1951. Courtesy Marshall Sawyer Jr.

Buddy Martello and fan dancer Sally Rand at the El Rey, ca. 1950. Courtesy Marshall Sawyer Jr.

Cottonwood Cove, fourteen miles east of Searchlight, ca. 1955. Courtesy J. Florian Mitchell.

The author's birthplace, following a winter storm, 1949. Courtesy Joyce Dickens Walker.

The author in front of his boyhood home in Searchlight, ca. 1945. The house is constructed of railroad ties from the Searchlight Railroad. Courtesy of Leland Sandquist.

Aerial photo of Searchlight taken on August 26, 1993. With permission of Cooper Aerial of Nevada.

Searchlight today, with Verlie Doing's Searchlight Nugget at center right. Courtesy of J. J. Balk.

23

THE CAMP WOULDN'T FAIL

The 1920s were the lowest point in the history of Searchlight. The population fell dramatically, with probably no more than a hundred people left in the camp. The state ordered the closure of the Quartette in 1922 when the station on the 650-foot level burned and two men died. This shaft did not open again until 1928.[1] There was very little commercial activity. Ella Knowles ran the general store, and a Mr. Black would bring food by stage that people preordered from him, much of the time on credit. The Cashmans still had their garage, at this time run by brother Harvey.

A study of the maps of the 1920s shows that between 1923 and 1927 Searchlight was on the way to Los Angeles and Salt Lake City. The Arrowhead Highway, the main Salt Lake–to–Los Angeles route, went through Searchlight. By 1927 U.S. 91 had been constructed, diverting most of the traffic away from Searchlight and further reducing its already limited tourism opportunities.

The notice that the great dam at Black Canyon was going to be built to harness the Colorado brought hope to the struggling town. This new project, to be called Boulder Dam, was less than thirty miles upriver. The act of Congress creating the dam was passed in 1928, and construction was completed in 1936.

Maude Frazier, the deputy superintendent of schools, stated in her 1921–22 report to the legislature: "Southern Nevada has a cause for its great optimism. Las Vegas will be the seat of operations, but the whole southern section will profit by the project. Cheap power will give impetus to mining."[2]

The rumor that the price of gold would be raised from its 1837 fixed figure of $26.67 an ounce stimulated new interest in the mines, mostly in leasing of the old properties. In 1934 the domestic ownership of gold was restricted because of the Depression and President Roosevelt's Emergency Banking Act. The action also set the price of gold at $35 an ounce.

The dam project allowed the men of Searchlight to find their first real work in many years. Housing in Las Vegas and the areas around the new dam site couldn't accommodate all of the activity that resulted from the construction, and so a market was created for Searchlight's amenities. The added travel induced by the dam also brought new customers over the dirt roads to Searchlight seeking gas and rooms, and old houses were rented to workers and visitors to the project.

One Searchlight resident described her quarters during the dam construction era: "We had an old shack. It was all we could find to live in. The windows were all out. When the wind blew, it would go creak, creak. We sat on orange crate boxes. An orange crate on the wall was our cupboard. A box spring and a mattress was our bed."[3]

During the 1930s Searchlight benefited from the influx of people that came with the dam project, as well as from the increased price of gold. The repeal of the Eighteenth Amendment also helped the town regain its personality of a wide-open, saloon-based western frontier community.

Carl Myers came to town in 1932,[4] obtaining one of the better jobs in Searchlight, at a salary of $5 a day. This was a significant improvement over his Depression-based wage of 34 cents an hour in Texas. In Searchlight, he worked at one of the still functioning shafts of the Quartette. At that time lessors Sid Gaines and Chris Kirkeby were working the Blossom with a great deal of vigor. A mill at the Pompeii site was still operating. Chief of the Hills, three miles below town, was digging several different tunnels in the mountains.

Here, as in other early mining towns, there was no concern about the environment. Mine waste and mill tailings were dumped and spilled without regard to the surroundings. For example, at McGill in White Pine County, where Kennecott had its mill, people were forced to move their homes to escape the lavalike flow of the tailings. In Searchlight there was no effort to contain the flow of the tailings. Today, cream-colored tailings ponds reveal evidence of old mill sites; even after ninety years, nothing grows there. Such mining practices would violate today's environmental protection laws.

Myers remembers that Ella Knowles ran the Searchlight Mercantile and Jim Warden operated the Desert Club, a local watering hole. Bert Mizer, the stepfather of piano player Rags M'Goo, operated the city water system, which got its water from a well in the middle of town. City water was very hard and bad-tasting, though, so most people hauled their water in barrels from the M&M Mine.

During the early 1930s, a doctor from the East moved to town because of his serious lung problems. There was not enough business for this physician, so he also became the postmaster of Searchlight for six months. Dr. Robert Fenlon would stay in Searchlight until the close of World War II. Bernice Kamer replaced him as postmaster, serving until the war ended. Kamer and Fenlon were married in 1935 and moved to Boulder City after the war.

Carl Myers recalls that when he first came to Searchlight, the Duplex Mine was operating only meagerly. Tailings were being milled, as were some of the old dumps, in the belief that the original mining and milling techniques did not allow for fully efficient processing and that newer techniques might discover previously hidden gold. Carl quickly learned about what everyone who lived in Searchlight was aware of: the widespread production of illegal liquor. There were not many empty buildings when Myers arrived, since fires over the years had destroyed a number of the permanent structures. In 1937, electricity finally arrived, from Boulder Dam. The construction crews for the power line from the dam to Los Angeles stayed in Searchlight for several months, significantly boosting the town's economy.

About the same time Myers arrived, Clara Bow and Rex Bell left Hollywood for the solitude of the desert, taking over a large section of the old Rock Springs cattle operation, which had since failed, and restoring it as a working ranch. Their base ranch facilities would be constructed seven miles west of Searchlight, in the middle of one of the largest Joshua tree forests in the world.

Rex Bell was recognized as the epitome of charisma by everyone who knew him. Over his career, he made thirty-seven B-grade western movies, none of them classics. His wife, Clara Bow, became a star of silent movies and remained a star into the age of talkies. Of all the famous movie stars over the years, none shone brighter than the "It Girl." After winning a national beauty and talent contest at age thirteen, she went directly to Hollywood. By the time she was seventeen, she was the grandest personality in the worldwide movie industry. These two stars met when both had roles in the same film; it was love at first sight, and they were quickly married.

Clara gradually developed medical and emotional problems, and the Bells retreated to the desert in an effort to protect her from the rigors of Hollywood. She forsook her film career to care for her two boys, who were born after the move. Rex ranched a little and socialized a great deal, eventually selling the ranch in 1950 to retired seaman Karl Weikel.

Bell's personality propelled him into Nevada politics; he was easily elected lieutenant governor in 1954 and was reelected four years later. All Nevada political writers of 1962 and thereafter believe Bell could have easily been elected governor had he had not been felled by a sudden heart attack in 1962 as he was campaigning for that office.

During the 1930s, Rex and Clara would sometimes come into town. They participated in some of the community activities and were always generous in allowing their names to be used in civic events, such as Independence Day celebrations. These two Hollywood personalities played a significant role in reintroducing Searchlight to Nevada. Clara Bow also contributed another unique personality to the folklore of Searchlight—her father, King Bow. Some recall that this short, rotund man attracted attention by walking about Searchlight's business district wearing only a bathing suit with no top and regular street shoes.[5] After Rex Bell's death, Clara faded even further into obscurity. She died in 1965 in a Southern California sanitarium.

One of their sons, Rex Bell Jr., became a judge in the Las Vegas area and was eventually elected district attorney of Clark County twice. He retired undefeated, one of Clark County's all-time most popular politicians.

One of the problems with the coming of the automobile was its intersection with the open range, where cattle roamed at will. The roads around Searchlight were driven carefully because of the free-roaming cattle. If a motorist hit an animal, laws in effect through the 1970s required the motorist to pay the cattleman the value of the animal. In 1939 the road from Las Vegas to Searchlight was paved, which compounded the problem—the cars then traveled faster but the cows didn't. The new road, however, made Searchlight more of a transportation hub.

With the advent of the 1930s, the town began a period of modest growth. The Yeager family came from Texas in 1931, started the Texas Cafe, and purchased the old Mine Operators Hospital, which still stands in Searchlight and is still owned by the Yeagers.

Jane Reid and her friend Helene Wilders opened the Monogram Cafe. Ray and Ella Shaeffer purchased Ella Knowles's mercantile and operated it. Their son later became the chief of police in Las Vegas.[6] A new service station was also built to meet the demands of the dam traffic. Surprisingly, in the early 1930s a Chinese restaurant operated by Doug Wong proved to be a popular spot.

With all the bootlegging that went on and the lack of law enforcement, the town saw a significant amount of violence. Some of the violence was self-inflicted: longtime resident and businessman Jess Thomas committed suicide, as did Jack Givens and Charles Hudgens.

One of the boys attending school in Searchlight at the time was a youngster named William Nellis. All class photos show his smiling face, seemingly overshadowing others in the photo. His grandmother, Olive, was the manager of the Searchlight Hotel and used to play music there. Once she was nearly beaten to death by an unruly customer. She was known to occasionally drink to excess, as did many in Searchlight.

Bill Nellis's story is remarkable. Bill's parents broke up when he was fourteen. He was left on his own, when his mother moved to California, and his father, Cecil Nellis, abandoned him and his sister. Cecil and his girlfriend, Dot, were town drunks. Cecil would work a claim for a few months, accumulate enough ore to be processed at Kelsey's mill, then go on a drunken spree until the money was gone.[7]

During high school Bill worked at many different menial jobs, including one with future U.S. senator Berkley Bunker, who owned a service station at the corner of Fremont and Fifth Streets in Las Vegas. Bunker remembered Bill Nellis as one of the finest young men he ever knew. He was small in stature, only five foot six or seven, but he was large in heart.

After playing football at Las Vegas High School and graduating in 1936, Bill married his childhood sweetheart, Shirley Fletcher, who was only fifteen at the time. Shortly after the wedding, they moved back to Searchlight and he worked for his uncle Jasper in the mines. He felt satisfied, since he was making good money at $5 per day. Searchlight was a good place to live because the rent was cheap and work was plentiful. As Berkley Bunker said, everyone who knew Bill liked him.[8]

While working at the Blossom for Kirkeby and Gaines, Bill was often called upon to drive Kirkeby in his big limousine to his old home in Ely, Nevada, a distance of about 300 miles.

Nellis finally got a steady job at the Union Pacific Railroad as a brakeman. This was a career with job security, considered essential during the war, so Bill did not have to fear the draft. One night as he and Shirley walked out of a movie, however, he said to her, "Should I join?" She responded, "Do what you think is best." He did. He was twenty-four years old, but he had already taken flying lessons before he entered the service on February 24, 1941. He was given the opportunity to go through flight training even though he was older than most of the other new pilots.

Shirley took their two babies (Gary, born in 1940, and Joyce, born in 1943) to see Lieutenant Nellis for the last time at his training base in Georgia. She drove from Nevada with her two infants so she and her childhood sweetheart could spend a few days together before he shipped out to the fields of war.[9]

He became a P-47 pilot and was assigned to the European theater in early 1944, when air activity was at a fever pitch. He flew seventy missions and performed many acts of heroism. On one mission his instruments failed. As he frantically tried to keep the aircraft from crashing, he tried desperately to get the canopy off so he could bail out. He rammed the canopy with his head until the canopy finally broke loose, and he could force his way through the small opening and bail out, even though he was blinded by blood from his head wounds. As he fell he lost consciousness. Ground troops found him hanging from a tree, his face and parachute covered with blood. He required many stitches but was soon back on the flightline.[10]

Another longtime Nevadan, businessman and county commissioner Richard Ronzone, served in the army during the war and was stationed at a base near Luxembourg where he was assigned to an anti-aircraft unit. He and Bill Nellis, who had been friends before the war, met by chance at this spot in Europe, so many miles from their homes in Nevada. By then Nellis had flown sixty-nine missions over enemy territory. He lamented to Ronzone that they would probably never see each other again.[11] Nellis had already flown enough missions to come home, but he volunteered for one last mission, since the unit was short one pilot. He did not return. He died during a close air support mission in the Battle of the Bulge, over Bastogne, Belgium, when his P-47 flew into heavy flak two days after Christmas, 1944. He was twenty-eight. His body was not recovered until April 1945.[12]

Nellis earned the Air Medal with silver and bronze oak leaf clusters. He was awarded numerous other medals, including the Purple Heart.

This boy from Searchlight was honored—as were his widow and children—when on May 20, 1950, the former Las Vegas Gunnery School was named Nellis Air Force Base in honor of the fallen hero. The nice young man with the deadbeat dad is now at rest in an American military cemetery in Belgium, near the small village of Henri-Chepelle.

When Ella Sandquist Kay came to Searchlight in 1934, the town was renewed with activity. There were no places to live, as most homes were burned or occupied. The only pastime she remembers during the 1930s was drinking liquor.

James Albert Leavitt maintained the dirt county road in Searchlight for a good part of the 1930s, a very difficult job. His equipment had solid rubber tires, and on a good day he could grade almost ten miles.[13] Leavitt died in 1995 at the age of 100.

During the 1930s, thanks to the higher price of gold, mining made a vigorous return to the Searchlight area. In 1931, 1935, and 1938, more than 3,000 ounces were produced; in 1936 and 1937, more than 5,000 ounces. These numbers seem impressive until they are compared with the first decade of the century, when production exceeded 10,000 ounces every year, with one year topping out at more than 25,000 ounces. Most of the gold, even in the 1930s, was produced at the Quartette.[14]

Bill Kelsey operated a homemade ten-stamp mill powered by electricity. The entire facility, built by Kelsey himself on the south front of Mount Doherty, started operations in 1934. It did not use commercial power because none was available; instead, Kelsey hooked up a diesel engine that powered a 3,300-volt generator. Diesel fuel was brought by truck from California, and water for the mill was imported from various places. Eventually the source was the Water Spout mining claim, located about where Verlie Doing now has her casino, the Searchlight Nugget. This little well produced so much water that Kelsey was able to sell the excess to others, including Rex Bell. Though some water was too bitter for the townspeople's taste, the cattle liked it.

Kelsey would assay the ore brought to him and then pay the miner 90 percent of the assayed value. Most of the ore came in very small loads; loads of only half a ton were not unusual. Because the mill was available, local miners could get instant money from their prospects, rather than having to wait weeks and sometimes months while the ore was shipped by truck to a rail stop, usually either Nipton or Boulder City, and then by rail to a distant smelter. Kelsey's operation was one of the best in the

West, as the mill was able to recover 98 percent of the value of the gold and silver in the ore. His customers came from all over the vicinity, some from as far away as Death Valley and Arizona.

Kelsey also worked the Quartette, not as a miller but as a miner. Kelsey and Esler Douglas, his father-in-law, explored one of the old shafts of the Quartette and took a few samples. One of them assayed at $30,000 per ton. By the time they got back, a cave-in had occurred at the shaft. They stayed underground anyway and were able to take out about a ton and a half of extremely high-grade ore in sacks. This small find netted Kelsey and Douglas $20,000 each—big money during those Depression years.

In 1940 Esler Douglas became the first person to be killed on the new paved road from Railroad Pass to Searchlight. He was forty-five years old, and his death shattered a remarkably successful partnership that had helped give life to Searchlight. Bill Kelsey continued to operate the mill until the war effort cut off his supplies. He closed the facility in 1942 and purchased a machine shop in Las Vegas.[15]

Gold was not considered essential to the war effort, and therefore precious-metal mining in the United States became a war casualty. The War Production Board's Gold Mine Closing Order L-208 and other official restrictions denied equipment, supplies, and even manpower to the gold and silver mines of the West.[16] The years 1940 and 1941 were fairly good production years for gold in Nevada, but without the necessary supplies for mining, mining lost and the war won. Searchlight's gold production fell to 897 ounces in 1942; in 1943, the figure dropped further, to only 139 ounces, and in 1944 to 21 ounces.

The war did benefit the economy of Searchlight in other ways, however. On March 4–7, 1942, General George Patton toured the area of southeastern California, southern Nevada, and southwestern Arizona that would be designated as the U.S. Army's Desert Training Center, 18,000 square miles in all. In an act of 1943 the name of the training region was changed to the California-Arizona Maneuver Area, and by late that year the total deployment stood at 160,000 men.[17] Camp Ibis, which was part of this army center, was only about thirty miles from Searchlight. Stationed at Ibis were upwards of 10,000 lonely men from all parts of America, there to be trained for the invasion of North Africa. This desert training camp was one of several in the huge complex.

One example will illustrate the economic impact that the arrival of the military had on Searchlight: Jane Reid, who owned the four-booth Monogram Cafe, could not cook enough food for the soldiers who came to town. They bought everything she had and more. Some were supplied three meals each day for the grand total of $1.25. Pies sold for $1, and she had trouble keeping up with the demand. "They would eat anything," she recalled.[18]

These soldiers also caused problems for the residents of the small town. One night seventeen soldiers were arrested and jammed into the small city jail. The next day the military police arrived, as usual, and hauled them back to the base brig. But before the military police arrived, friends of the soldiers attempted to tear the jail down. They failed, but it did make for an exciting night.[19]

Tonopah experienced a similar situation during the war. Thousands of army airmen were trained at an army air base about eight miles from town, and their presence was a boon to Tonopah's economy. Both military operations were closed down as the war ended. Little remained of Camp Ibis, but the Tonopah air base was turned over to civilian control and is still in use today. Now, after fifty years, the desert has been cleared of the tracks of tanks and other evidences of war.

24

THE RENEGADE INDIAN

On February 21, 1940, the banner headline in the *Las Vegas Review-Journal*—BODY OF INDIAN FOUND—recalled for many in the town memories of the first murder the dead Indian had committed, thirty years earlier at Timber Mountain, just a few miles from Searchlight in the McCullough Range.

On a cool fall day in October 1910 Harriett and John Reid were on their way, via horse-drawn wagon, to work at their mine—she manned the horse-operated hoist, he mined the ore. They could see an Indian approaching them, carrying a .30-.30 Winchester rifle and traveling at a very fast pace. The Reids stopped, as did the Indian, whom they recognized as Queho, an acquaintance who worked at various menial jobs throughout the Searchlight area. They exchanged greetings and after a brief visit went their separate ways. Later, the Reids and everyone else in the area learned that Queho had been hurrying down from Timber Mountain, where he had been cutting wood for J. M. Woodworth, a timber and firewood contractor. Woodworth had refused to pay Queho, who then flew into a rage and beat the man to death with one of the timbers he had cut. This murder was the beginning of an odyssey that took thirty years to play out.

Queho soon struck again, this time near the river in Eldorado Canyon, at the Gold Bug Mine, which was partially owned by Frank Rockefeller, brother of John D. Rockefeller. A short time afterward, Queho admitted to Canyon Charlie, an Indian elder almost a hundred years old, that he had killed the mine's night watchman, his former employer. The second murder occurred on the route between the Crescent area, where the woodcutter was killed, and the river.[1]

Local lawmen, who viewed Queho as little more than an ignorant savage, thought that catching him would be child's play. They couldn't have been more wrong. The clever Indian stole a horse from a man named Cox and eluded the law.[2]

A large manhunt was organized to apprehend the Indian outlaw. It was assumed that Queho would be easy to track, since he dragged one leg as a result of an earlier injury. James Babcock, an operator of the Eldorado Mine and a lawyer educated in Washington, D.C., led the search party. He was accompanied by a contingent of Las Vegas lawmen, including Ike Alcock, as well as Indian trackers and an Indian agent named DeCrevecoeur. One of the pursuers was overheard remarking that Queho's chances of living a long and happy life were very slim.[3] The manhunt extended more than 200 miles, ranging from Crescent to Nipton and even coursing toward Pahranagat Valley, nearly 150 miles to the north. The pursuers gave up the search when supplies ran out and they grew weary. At that point the lawmen began to suspect that maybe this Indian was cunning and smart, not quite the "dumb" savage they had thought.[4]

Queho was subsequently blamed for a number of murders that he did not commit. The first was the murder of James Patterson. The newspaper headline read, MAN KILLED BY QUEHO STILL ALIVE. Patterson hadn't been killed by the Indian or anyone else—as was evident when he turned up alive and unharmed. But in the course of looking for Patterson, the search party found another man whom Queho had shot.[5]

The press closely followed Queho's escapades. A reward of $500 was offered for the Indian's capture, and Nevada's only member of Congress announced that the federal government should assist in the capture of this madman.[6]

In March 1911 it was reported that two men on the Arizona side of the river, just below Searchlight, watched Queho beat a white man to death on the opposite side. The prospectors were powerless to help, as they had no way to cross the river; they were also unarmed and feared that Queho was armed and would attack them. By this time fear gripped the entire region.

It was believed that the best method for apprehending Queho was to enlist the Piutes in the search, which was standard operating procedure at the time.[7] Whites regularly abused and harassed the Indians, and if an Indian committed a crime, the white community would force the Indians to produce someone to answer for the crime. To fail in this responsibility meant great distress for the Indians because it led to further harassment by the whites.

In the hills below Searchlight, about five miles from the river, one of the Du Pont heirs to the chemical fortune of the eastern United States was encamped. He was an outcast from his famous family. At the urging of voices that only he could hear, he had begun digging a tunnel through one of the volcanic mountains with a pick and shovel. He started the tunnel in 1896, even before gold was discovered in Searchlight, and eventually extended it nearly 2,000 feet through the solid volcanic rock.[8] Du Pont was always friendly to the Indians who came by his camp and often shared his provisions with them. But shortly after the murders of Woodworth and the Gold Bug watchman, some of Du Pont's supplies disappeared, and Queho was said to be the culprit. The newspaper editorialized that the federal government owed a responsibility to the people of Searchlight to intercede in this Indian affair. It wrote: "A good Indian is a dead Indian."[9]

Most still believed that Queho would be caught, that with both Indians and whites on his trail victory was assured. The *Las Vegas Age* newspaper headlined an article with QUEHO THE BAD INDIAN IS IN A BAD FIX. In a subsequent edition the paper said that civilization and bad whiskey had turned Queho into the killer he was. The paper also observed: "It is very probable that Mr. Queho's days are numbered considering those after him."[10]

The posse was large and well equipped, as all the other hunting and tracking parties had been. At this point it was believed that Queho had come back to the river. Alcock wrote to Constable Colton in Searchlight, informing him that he was on the trail of Queho, as he had recently found fresh tracks at Cow Wells, near Searchlight. Queho was also reported to have been seen in the town itself at least once.[11] The posse came up empty-handed.

In 1912 Fred Pine, while hunting near Timber Mountain, came upon Queho, who was armed with his ever-present Winchester. The men exchanged greetings. Pine asked Queho if he would like one of his sandwiches. Queho accepted and in return offered Pine one of his dried rats or chipmunks. Pine finally turned to leave, expecting at any moment to be shot in the back, but nothing happened.[12]

Queho was surely an expert at hunting and fishing. He could eat anything, including tortoises, chuckawallas, burros, horses, mountain sheep, chipmunks, rats, and various birds.[13]

The Queho legend began to grow. Several manhunts were organized—all public, all ending in failure. The Searchlight newspaper ceased publication, so news about the comings and goings of the fugitive was no longer so sensationalized. Though some believed he had been killed by other Indians, occasional sightings were reported. There were even rumors that he had a girlfriend around Searchlight named Indian Mary. Others reported having seen him in Searchlight.[14] Murl Emery told people that he had seen Queho several times. Searchlight residents indicated that some contact was maintained with him over the next twenty-five years.

Seven years later, in the winter of 1919, the peace of the countryside was again shattered when Maude Douglas was murdered in her home at the Techatticup Mine in Eldorado Canyon. She heard a noise in the dead of night, walked into the kitchen to investigate, and was felled by a shotgun blast. On the floor was spilled cornmeal that the intruder had been trying to take from the cupboard. The trail from the cabin showed tracks of a man with a noticeable limp, like Queho's.[15]

Mrs. Douglas was married and had two children of her own, as well as responsibility for two other small youngsters, Bertha and Leo Kennedy. Leo, who was only four years old at the time of the murder, later said that Maude had been killed by Arvin Douglas, the man of the house.[16] There is no corroborating evidence to support that claim, especially in view of the uniquely patterned tracks at the Douglas cabin. Bertha also said that she felt responsible, because she had awakened Mrs. Douglas for a drink of water and if she had not done that, the woman would not have gone into the kitchen.[17] The overwhelming weight of the evidence pointed to Queho, as confirmed by a coroner's inquest that was convened after Maude's death. The coroner determined that she had been shot at close range and that the tracks from the house fit Queho's.

The murder of Maude Douglas initiated a new era of Queho hunting. During the chase, the search party found a mountain sheep that Queho had recently slaughtered. They also found two dead miners named Taylor and Hancock, whom he had killed with their own prospector's pick. The searchers soon learned that Queho traveled at night and holed up during the day. The pursuit ended in futility after three weeks, with the near death of the group's leader, Frank Wait, from exhaustion.[18]

Wait believed that Queho was hiding in the area where he had killed Woodworth. Knowing that he was being followed, Queho did not want to attract attention with gunshots, so he killed the two miners with their pick, probably to get a replacement for his worn boots. Sheriff Joe Keate described him as being able to starve a coyote to death and still have plenty of strength to continue. He reportedly knew of places in the desert where depressions worn into the rock stored rainwater for up to a year.

Alcock, a man named Alvord, an Indian trader named Baboon, and ten others made up the search party. Among the group were some Indians, and it was discovered that they were signaling Queho by smoke signal, thus allowing the killer to elude his pursuers.[19]

The reward was increased to $3,000. Individuals and groups found evidence of Queho—a cave he had stayed in along the Colorado, remains of a mountain sheep and a burro.[20]

For the next few years, another period of quiet prevailed when no recorded murders were known to have been committed by Queho. Nevertheless, no one felt secure. Prospectors and others tried to travel in pairs, one or the other of them always keeping watch at night. Not until 1935 did the next confirmed sighting of Queho take place. A cowboy named Charles Parker had a mare disappear; a week later the horse was found with part of its carcass cut away, obviously for eating. Upon investigating, the cowboy got more than he had bargained for. He was accosted by a scantily clad Indian with long, stringy hair and was robbed, but escaped unharmed. Searching the same area later, Parker and others found a cave along the river with drying jerky in it. A gunfight ensued and nine shots were fired, with no apparent injury to either of the parties.[21]

As the years passed, Queho was accused of killing as many as twenty-one people. His first murder actually occurred before the Woodworth episode; the victim was his cousin or half brother, an Indian outlaw named Avote. The white community insisted that the Indians produce someone to pay for Avote's crimes, and so as a young man, Queho killed his relative at Cottonwood Island on the river below Searchlight. He also likely killed Bismark, a Las Vegas Indian, but that was a tribal killing and would not usually have been pursued in early Las Vegas. There were allegations of other killings but no actual proof.

Queho outsmarted the best that law enforcement had to offer. His pursuers may have come close on several occasions, but he always evaded them. He was an excellent shot and had a reputation of being extremely brutal.

Finally, in February 1940, Queho's body was found by three prospectors in a cave about ten miles below Boulder Dam and 2,000 feet above the river.[22] They also found fuses and blasting caps from the dam at the site. This cave was one of the best hidden and most impregnable hideaways imaginable. It even had a trip wire hooked to a bell to alert him of intruders. Queho had been dead for at least six months.

Some of his old pursuers, not wanting to acknowledge that they had been outsmarted for thirty years, tried to say he had been dead since

1919. Items in the cave from the construction of Boulder Dam quickly disproved their claim—veneer board, used in concrete moldings at the dam, that Queho used for protection from the elements. And there were fuses, which he used for reloading his bullets and shotgun shells. Also discovered in the cave was the badge of the night watchman killed at the Gold Bug Mine. His loaded Winchester rifle and the shotgun with which he likely killed Maude Douglas were in the cave, as well as a fine bow and twelve steel-tipped arrows (probably for fishing in the river), recently minted coins, and papers from some of his victims.

The large number of eyeglasses in the cave probably indicated that he was afflicted with poor eyesight in his later years. At death he was believed to be about sixty years old. He had died in a position of apparent pain, wearing a canvas hat and pants.[23] One of his legs was wrapped with burlap, which indicated that he may have been snakebitten. A former acquaintance confirmed the identity of the body by the unusual dental feature of double rows of teeth.[24]

Charley Kenyon, one of the prospectors who discovered Queho's body, later found other nearby caves that the Indian had used. Queho was also said to have panned a little gold, which he saved in Bull Durham tobacco sacks, then exchanged for food and other supplies. One of the persons who probably had some contact with Queho was the eminent Murl Emery, who always seemed protective of him and also admitted to leaving food for him. Emery was quoted as saying, "Why don't you let the poor Indian rest?" Emery lived at and operated Nelson's Landing for many years and was a constant companion of mystery writer Erle Stanley Gardner.[25]

Queho remained controversial even after death. Two political enemies and former law enforcement officers, Gene Ward and Frank Wait, both involved in trying to bring the desperado to justice over the years, fought over his skeleton.

Neither won, as James Cashman and the Elks Lodge intervened to pay the funeral home for the costs of interment. The Elks then displayed Queho's bones at Helldorado (the premier entertainment event in Las Vegas for more than forty years, beginning in the 1930s) as a carnival attraction. The bones were stolen from Helldorado Village and found in

Bonanza Wash in Las Vegas; Dick Seneker subsequently acquired them and returned them when James Cashman again offered a reward. The Indian's remains are now believed to be buried in Cathedral Canyon near Pahrump.

Queho's name continues to bring forth tales too numerous to confirm. In an oral statement taken in the late 1970s, historian Elbert Edwards of Boulder City gave a rambling account of stories about Queho. Edwards did not rebut the stories of Queho's murderous binge, attributing a total of seventeen murders to the Indian.

Edwards described one man who was killed with a pick handle before Woodworth was killed at Timber Mountain. He then confirmed the murders of the Gold Bug watchman, Maude Douglas at Nelson, the two St. Thomas miners Hancock and Taylor, and then two unidentified miners. He also described Queho's murder of a wandering cowboy with his trusty rifle and spoke of five individuals who were killed in a cabin near what is now Boulder Dam—three with a rifle and two with a knife. Edwards's narrative also related the story of two others, killed in nearby Black Canyon the next day.[26] The authenticity of most of the murders recounted by Edwards is questionable, but they do reveal the legendary status accorded this Indian desperado.

Queho was a killer who outsmarted all who tried to capture him. The story is tragic, not only because of the lives that he took but because even in Searchlight his story illustrates to us how poorly Indians were treated. The first census, in 1900, reported forty-two Indians in Searchlight, obviously in the river area where there was water. They were eventually driven out of Searchlight.

In the summer of 1905 the Searchlight newspaper reported that the Indian village on the outskirts of town had been destroyed by fire. The paper disparagingly remarked: "All bucks and squaws were away."[27] Indians were granted no respect in Searchlight, and they were harassed and discriminated against in increasingly offensive ways. It is no wonder that Queho's fellow Indians helped him. Nor is it surprising that he became known among the few Indians of the area as someone who had stood up to the white man.

25

SEARCHLIGHT SURVIVED THE WAR

Even as the discovery of Queho's body in the rugged terrain on the river between the dam and Nelson's Landing brought back memories of the Searchlight of days gone by, the town was being thrust into a new era of its history.

Some of the old-timers were hanging on, as exemplified by Al Marshall, who in 1942 drove 150 head of cattle from Kingman south to Needles, then over the river and north to Rex Bell's ranch, a distance of more than 100 miles, from one arid desert area to another. At the time, mining ventures were being squeezed out of the old mining camp, and the Searchlight school board found itself without funds to pay the teacher for three months. When all else failed, the townspeople held bingo games in the saloons to raise money for the teacher's salary.[1]

When the war ended, people could more easily travel, and that helped Searchlight's small business community. There were lots of cars on the road, especially after plans for the construction of Davis Dam, just twenty-five miles away, were announced. A town just across from the dam site, Bullhead City, Arizona, became headquarters for the construction and subsequently the dam itself. Bullhead City already existed before the dam, but the Nevada side of the river was undeveloped until after the dam construction.

The mines, which had been virtually closed during the war, found it difficult to reopen. The Quartette, the mother lode of Searchlight, was in disrepair, and the main shaft was impassable.

With the spirit of free enterprise kindled anew, unique businesses arose in Searchlight. John Kay came to town and started a brickmaking operation, which lasted about four years. Howard Mildren began the construction of plywood boats, a business that lasted about three years. There were other speculative attempts at original businesses, such as using yucca and Joshua tree fiber to make various products.[2]

In the fall of 1947 five liquor licenses for Searchlight business were granted at just one county commission meeting, and several others were tabled because a county ordinance required the expenditure of $3,000 before a license could be granted. Why the activity? Construction of Davis Dam, continued military activity at the Las Vegas Gunnery School (soon to be renamed Nellis Air Force Base) and at Indian Springs Air Force Base and Camp Desert Rock, plus the recently initiated activity at the Nevada Test Site stimulated long-standing prostitution operations in Searchlight. Government officials in Las Vegas were even talking of building a new jail to handle the added criminal activity that accompanied the promiscuous business in Searchlight.[3]

For a couple of years beginning in 1946, a weekly tabloid newspaper was published with the masthead the *Searchlight Journal*. As early as the October 31 issue, this publication listed several of the businesses then operating: Oasis Club, Golden Eagle Bar, Desert Club, Crystal Club, Searchlight Casino, El Rey Club, Union Bar, Hotel Rondo, Workmans Club, and Nevada Hotel. But in addition to the bars and houses of prostitution were Ed's Cafe, Chapman Grocery, Union 76, McDonald Electric, Al's Trailer Court, Chevron Gas, and the Searchlight Surplus.[4]

The Searchlight Little Theater produced only one show, a play sponsored by the Desert Club. The show closed after its opening night, because the attendance was less than fifteen people. This old camp wasn't interested in the fine arts.

Political candidates, both Democrats and Republicans, continued to come to town to market themselves in seeking support for county and even statewide office.[5]

A brand-new school was constructed in 1942, the land donated by Dewey Yeager to clear title to the location where the school had long been located. As reported earlier, the old school was moved to a site on the Las Vegas Highway about half a mile from the old site. The building was used as a bar, but it burned to the ground shortly after the move.[6]

Fire revisited Searchlight in August 1946, leveling Shaeffers Grocery, the Miners Club, and two other structures. These businesses never reopened.

A new city hall was constructed just below the school in a war surplus building donated by John Kay. John and his wife, Molly, also built the Kay Motel from war surplus buildings. By most standards this seventeen-unit rural motel was as modern and functional as any in Nevada.

Mining in Searchlight picked up again, with at least two lessors working the old Quartette property, and the Blossom. The Duplex had a flirtation with production, as it had on many occasions since 1910, when Bill Roller, a relative of the Colton family, hired twelve men in 1948 to work the mine; the project lasted just over a year.[7]

The building of modern homes, especially so many of them in Southern California, created a large demand for nonmetallic minerals. Limestone and gypsum were desperately needed, and both were readily available in Nevada. Perlite, a substance that functioned somewhat like gypsum and was used primarily for insulation and wallboard, was found at many Nevada sites, including Lovelock, Reno, and Searchlight.[8] The new perlite operation in Searchlight started in 1947, after the property was located in 1946, about four miles northwest of town.

This reserve of perlite was estimated at more than 100 million tons. The mine and mill went into operation in 1947 and continued until the owner, Byron Tanner, died in the early 1980s. Three men ran both the mill and the mine. They would mine enough in the open pit to run the mill, and then all three would return to the mill to service the ore. Mining and milling operations were conducted separately; when one was in operation, the other would be silent. The small Searchlight operation supplied materials for a number of hotels in Las Vegas, including the Thunderbird and El Mirador.[9] After Byron's death, his wife, Elaine, ran the operation herself. She was, in her early days in Searchlight, a striking brunette. With her long coal-black hair, she made a real impression when she came to town. In spite of her beauty and hard work, her small operation could not compete with U.S. Gypsum, Johns-Manville, and the other large suppliers of products similar to hers.

By 1947 the town board, an entity established by the Clark County Commission, chose two of its residents to go to Carson City to lobby the legislature to pave the Nipton Road to make travel to Searchlight easier. This project had been on the state road construction agenda since 1939.[10]

The Las Vegas newspaper during this period took note of the fact that Rose McDonald, nearing her eightieth year, had come to town in her old car, traveling from Spirit Mountain, fifteen miles over very poor dirt roads, to get her monthly supplies. McDonald was well known. In the mid-1930s, shortly after she was widowed, two men worked her mining claim without her permission, stealing her ore. When she asked them to leave, they responded, "No, woman." She went home, came back to the mine with her rifle, and killed both of the claim jumpers, then loaded the bodies into her car and brought them into Searchlight, where she was arrested. Her trial for murder was the first held in Las Vegas's new courthouse. She was acquitted.

The newspaper business in Searchlight had never been a successful enterprise, as Scoop Mildren would learn when he began publishing the *Searchlight Journal,* a local paper that focused on local news but also provided national columns. This paper ended in 1948 with Mildren proclaiming, "Our aim from the very start was to make a small town family newspaper, not a scandal sheet. When Big John threw Joe Deaken in the can to sober him up, when bloody noses and butcher knives were part of the week's celebration, the Journal closed its eyes and looked for mining news. This is not the way of a newsman. It irks him to save the black eyes and pour on too much honey. It also irks and hurts him to send out bills and receive no pay. It also hurts him when people withdraw their ads because they have a personal dislike for his partner—especially when, who his partner is, in fact, has nothing to do with the value his paper gives in publicity for the town. For these and several other reasons then, with this issue, the Journal is taking a vacation."

Mildren was one of many characters in Searchlight. Another favorite was Pete Domitrovich, better known as Big-Nosed Pete. He was unusually strong, able to carry two railroad ties at once, one under each arm. He could put a loaded fifty-gallon drum on the back of a truck. One night in the late 1950s, as he was leaving Sandy's Bar, he observed two men siphoning gas out of a car parked near the front of the bar. He proceeded to call them "dirty gas stealing sons of bitches" and beat them both into submission. When the dust settled, Pete learned that the car from which they were siphoning was their own.

Another character was a man known only as "Monk." Miner's consumption was common among the miners of Nevada. Some places were worse than others, and Tonopah was the worst, while some of the mines in Searchlight were bad also. Monk was a classic example of a miner afflicted with the disease. When he was outside and went into one of his coughing fits, his hacking could be heard from one end of Searchlight to the other.

Another legendary figure was Charlie Johnson. He was born and married in Sweden. He traveled to Mexico and finally came to Searchlight as a hand on the railroad. He was known for the story of the camp rooster who crowed too loudly. One night when the crew had drunk too much and did not want to be awakened early, they force-fed lard to the rooster. Charlie loved to crow like the rooster with lard in his throat.[11]

When Charlie died, Lonzo Dickens from Searchlight handmade his coffin. People at the local bar donated enough money to take his body to the Palm Funeral Parlor in Las Vegas to be prepared for burial. Charlie was taken from Searchlight in his homemade casket in the back of a pickup by several friends and mourners. Before reaching Las Vegas, the mourners got thirsty, so they stopped for drinks at the Railroad Pass Bar and Casino. For years afterward, the bartender laughed about the Searchlight crowd drinking to old Charlie Johnson while he was parked just outside in his casket.

Though the veracity of these stories might be called into question, the fact remains that Searchlight had its share of unusual personalities.

26

THE PROMOTER

From the very beginning, Searchlight was home to some innovative entrepreneurs. Colonel Hopkins, the early successful owner and operator of the Quartette, was the epitome of a good businessman. The years that followed brought others to Searchlight who made the most out of some very limited opportunities.

The Searchlight gold mines had been almost silent during World War II. Other mines in Nevada flourished during the war because they produced minerals used in the war effort. Copper, lead, zinc, tungsten, and manganese were heavily mined; in 1952 Nevada had sixty-four tungsten mines.[1] Except for copper, none of these minerals were mined extensively after 1955.

After the war, a savior named Homer C. Mills appeared in Searchlight. With the apparent backing of a huge bankroll, Mills obtained leases on the Quartette, the Blossom, and a few other mines. His companies had impressive names like Golden Dawn, Homestake, Searchlight Uranium, Dutch Oven, and American Gold Mining and Milling; they were companies that many thought would bring back the days of Searchlight's early fame.

Mills moved into the old Quartette superintendent's home, a residence that harkened back to the days of the Quartette's boom years. The home, which sat on a hill, was best described as a rambling ranch-style house. It had eight bedrooms and was easily the nicest home in all of Searchlight, even though it was then old and somewhat dilapidated.

Mills also impressed the people of Searchlight because he had an office in Los Angeles and he always drove a big passenger car. In short, he was the type of entrepreneur that this poor town had been looking for since the mining boom had ended forty years earlier—an operator, a promoter, someone who would restore the real Searchlight.

He employed local miners, the first real hiring since the war. Never did he hire many, probably a dozen at the most. But he paid top money. Either he would pay straight wages or he would contract work, paying the miners $3 per foot in a drift or crosscut, more if it was a shaft.

What the miners and other mine workers found fascinating was that many people came to visit Mills at the mines. These visitors would be given a tour of the workings of the mines. As everyone soon learned, Mills was a salesman of the highest caliber.

When he first came to town in 1946, Homer Mills was fifty-nine years of age, standing about five feet, ten inches, and weighing about 180 pounds. He had long white hair and usually wore a cowboy hat and boots. He was a very handsome man, very articulate, and he spoke like a lawyer. He was known to be kind and gentle and even took one of the local elementary school kids with him to California on several of his trips.

The townspeople noticed, however, that Mills had begun to bounce lots of checks, especially those written to his hardworking employees. His word was soon not his bond, even though he seemed to be very gracious.

It was learned that the visitors to the mines were not sightseers but potential investors whom Mills had lured to town for the purpose of selling them stock in one of his companies, most of which were not even legally incorporated. Even though Searchlight in its heyday had been a producer of gold and silver, Mills told the unwary that the earth in Searchlight also contained vast treasures of lead, zinc, and even uranium. These claims were baseless.

To underscore his sales pitch, Mills would produce reports from experts who confidently stated the potential of the Searchlight ground. The accomplice he used most often was a man named Harvey Sill, who masqueraded as an engineer. This Los Angeles "engineer" issued a November 1947 report consisting of five written pages accompanied by ten pages of charts and graphs. The report "verified" Mills's claim that gold, copper, lead, and silver were present in abundance in the Quartette. It is clear that the document helped him sell stock in his enterprises. Even though Mills had leases on several properties, he usually worked only one mine at a time. Thus he could focus his attention on a single promotion without distraction.

In a subsequent report on property near the old Blossom Mine, Sill became much more bold. In this 1948 paper he stated: "These claims have an aggregate production estimated in excess of $5,000,000. These properties have been worked to depth and the Quartette has one shaft down to the 1300 foot level and was reported to be still in ore when shut down. I understand that no shaft in the Searchlight district has ever been bottomed for the want of ore."

It was easy to see that Mills was loose with the facts. But his salesmanship allowed him to make huge amounts of money from his misconduct.[2]

Homer C. Mills was born in Kentucky on April 10, 1887. It is unknown exactly when he came west or how he wound up in Idaho, where he was admitted to the Idaho bar at age twenty-three in 1910. He was able to sit for the Idaho bar because he was befriended by an Idaho attorney who saw promise in his young protégé. Homer Mills was of well above average intelligence, and he saw a way out of the poverty he had known all his life.

Before the 1930s an individual did not need to be a college or law school graduate to become a lawyer. A practicing attorney could sponsor a candidate to take the bar. Mills worked hard and eventually was permitted to practice law in Idaho. His talents for the law were so advanced that he applied to sit for the California bar and, in typical Mills fashion, was successful. He was an assistant U.S. attorney for Idaho during 1910 and 1911, and he became the elected prosecutor for Minidoka County, Idaho, in 1915.

It is not known for what reasons or at what precise time Mills moved to California; he was probably motivated by the economic potential for a lawyer in a rapidly growing state like California compared with the unspectacular prospects in the agrarian, very slowly growing state of Idaho. It is known that by 1925 Homer C. Mills was profitably engaged in the practice of law in California, because in March 1926 an agreement he entered into as an attorney became the subject of a California Supreme Court opinion.

This opinion gives insight into Mills's later deceitful practices in promoting Searchlight. In its opinion the California Supreme Court explained how Mills entered into an agreement in writing to represent one E. E. Easton in a series of legal problems, including a divorce proceeding.[3]

To put Mills's dealings in perspective, the court explained how he stole money from his client, padded his bills, improperly accounted for time and money, used a nonlawyer to render legal services, and took advantage of his position of trust. Mills's ingenuity as a legal tactician was evident in another of Easton's cases, which went all the way to the U.S. Supreme Court. In this 1931 decision the California court ruled: "Notwithstanding the serious misconduct of petitioner, we hesitate to inflict upon him permanent disbarment. . . . Moreover, petitioner with apparent skill and ability conducted a vast amount of litigation for complainant. He [Mills] is advanced in life and . . . may even yet, if given the opportunity, redeem himself from the odium of this present offense." The court then proceeded to suspend Homer Mills for one year from the practice of law. The U.S. Supreme Court did not review the case.

Unfortunately, Mills did not learn his lesson, for in 1936 he was disbarred from the practice of law in California.[4] In this proceeding the court explained again how he had violated his position of trust by taking money for his own use that rightfully belonged to his clients. Along with making other points, the court observed: "Five years ago this same petitioner was before the State Bar and this court, wherein it became necessary to suspend for one year his license to practice law. It is appropriate to take into consideration this previous record in deciding what penalty should be imposed for the misconduct involved in the present proceeding. Whatever might be said in favor of a milder punishment for single misdemeanors not involving deliberate moral turpitude, there is no doubt in our minds that any lawyer who is guilty of habitual misuse of the funds of his clients should be deprived of the license under which he is authorized to practice law, and by which he has been recommended to the public as a person worthy of trust. It is therefore ordered that Homer C. Mills be and he hereby is disbarred from the practice of law in this state and his name is stricken from the roll of attorneys."

Mr. Mills was a slow learner about the consequences of criminal activity. Not only had he committed crimes that led to his disbarment but he continued his criminal conduct and was subsequently convicted of a felony in the Superior Court in and for the County of Los Angeles on February 5, 1937.[5] These charges stemmed from writing checks with insufficient funds. He was sentenced to San Quentin Penitentiary and was incarcerated from November 20, 1937, to May 22, 1939. He received a pardon in November 1940.

It is not known what Mills did during the war. But it is known that he arrived in Searchlight in 1946. Why Searchlight? It is probable that for Mills, southern Nevada was close enough to southern California to allow him to exploit his solid client base for his future promotions.

While in Searchlight, he put his legal training to use on many occasions. The Securities and Exchange Commission began a long investigation of his stock dealings, focusing on the phony sales he had made in the Chicago area for stock he sold relative to Searchlight mines. A criminal complaint was filed charging him with eight counts of mail fraud and violation of SEC rules and regulations.

At the trial, where Mills acted as his own attorney, the government brought in several witnesses to prove his misrepresentations. The evidence established that he had taken nearly $300,000 from thirty-nine investors.[6] With daring chutzpah, Homer continued selling his stock in the Searchlight area during the entire criminal proceeding.

During the trial, an engineer testified that he had bought more than $1,000 worth of uranium stocks but could never get Mills to answer his letters or communicate with him in any way. An attractive widow invested in some of Mills's stock; she, of course, lost everything. A successful businessman was convinced by the gentleman from Searchlight that the Blossom had more than $4 million worth of uranium in its depths, a pretty good claim for a mine that never produced an ounce of the stuff.

The trial continued for more than a month, ending with Mills's summation to the jury, which was based on Adam Smith's *Wealth of Nations,* the 1776 publication that is the foundation for the U.S. system of free enterprise. In short, Mills told the jury of six men and six women that if the investors wanted to put their money in speculative ventures, then they should have the right to do so. The jury deliberated for only a short time before returning a verdict of not guilty. It is said that the jurors were so impressed with Mills's presentation that several invested in his schemes after the verdict was announced.

During all of this time the people of Searchlight knew nothing of his problems, but they did notice that more of his bills were left unpaid, including wages to his employees. One man sold him some mining equipment on which he defaulted. When confronted by the upset vendor, Mills talked him into a long-term lease, and the meeting ended on a happy note. It took some time for the man to realize that Mills had put one over on him again because it is much harder to collect on a lease in default than on a contract of sale. No matter how one dealt with Mills, he was bound to lose.

Mills was also a genius at checkers and chess, which he probably learned while in San Quentin. One Searchlight resident recalled a legendary incident when Homer challenged him to a checkers game using eight fewer checkers than his opponent had. Mills won the game.[7]

Homer's mode of operation was unique. He spent enough time in Searchlight to oversee the mining operations. But since his biggest concern was stock promotion, not mining, his time in Searchlight was secondary to his stock sales, which were for cash, not for gold.

Because he had so many bad checks floating around the small town, he did most of his entertaining in Las Vegas, where he had a reputation of being a big spender, especially at the gambling tables, where he is remembered for losing, not winning.

His ability to lead a life of profligate spending came from his never-ending sales prospects. Mills ran inexpensive advertisements in newspapers all around Southern California. The clients who purchased stock came from a never-ending supply of fresh meat for the ever aggressive lawyer and stock salesman. He kept a log of more than 500 potential investors, obtained from his many and varied contacts.

In Nevada gambling joints, hired shills are used to play at the table games in an effort to draw others to play in the same game. They do not get to keep their winnings and are covered in their losses by their employer. Homer Mills used this same method in selling his stock. He had various people who would steer clients to him for a small fee; they were his stock shills.

He was married and divorced twice, but the people around Searchlight got used to the many attractive women who accompanied him, some purporting to be his secretary.

One of his most successful selling jobs was to a man and wife named Hull. After completing the sale, Mr. Hull handed Mills a business card. The man was in reality Lieutenant William C. Hull, Bunco Squad, Los Angeles Police Department. Hull's wife was a fellow detective. This was the opening act of a fourteen-count indictment in the Superior Court of California. Twelve of the counts were for grand theft, one for attempted grand theft, and one for securities fraud.

When the arrest in California was made, Hull had with him two securities investigators, two patrolmen, and two police photographers. A search warrant was executed, and the resulting evidence produced at trial seemed overwhelming. Over the objections of the prosecution, Mills was allowed to remain free on bail from the time of his arrest to the time of trial. He occupied his time by selling more stock.

At the trial, of course, Mills acted as his own lawyer. The evidence depicted Homer C. Mills as a smooth, charming man who sweet-talked his way through life, swindling hundreds and hundreds of wealthy investors as well as middle- and lower-middle-class people who could ill afford the risk. Homer Mills did not discriminate; he would swindle anyone.

One schoolteacher gave Mills money on five different occasions during an eight-year period, a total of $14,000. Testimony showed that one fashion designer steered more than $100,000 of business to him from her friends and business associates. A physician from Watsonville bought $70,000 worth of securities personally but refused to testify, as he still believed Mills to be an honorable man.

At the conclusion of an extended trial, Mills used the same defense that he had employed years before in Chicago. After a lengthy deliberation, the jury returned guilty verdicts on all fourteen counts. Again, the judge allowed Mills to remain free on bail, pending appeal. So, Mills continued unhindered to sell stock on Searchlight mining ventures.

The appellate court reversed the guilty verdicts, on the basis of a case it had decided a short time before the Mills appeal was considered. In fact, the opinion was being written while Mills's trial was being conducted. This opinion was based on an improper search warrant and improper obtainment of evidence, including entrapment and illegal wiretapping. Homer C. Mills was free and immediately returned to his mansion on the hill at the old Quartette in Searchlight.

The Los Angeles district attorney was furious that the notorious con artist had beaten him, especially after Mills was heard bragging about the decision outside the courtroom. Mills should have laid low, since the DA's office was out to get him. However, the words "lay low" were not in his vocabulary.

It took only two months for another undercover police officer to arrange a meeting to consummate a fraudulent stock sale. The transaction took place in a Hollywood restaurant. Again, Mills was arrested and shortly thereafter was tried for grand theft and stock fraud. At this point his world fell apart. His age and years of hard living had taken their toll; lacking his usual vigor and originality, Mills was convicted. He continued to sell stock, this time to prison personnel, but in just a short eight months he died—without regrets, it was said. Searchlight's all-time greatest promoter had died, perhaps as the last Searchlighter of the twentieth century to advocate pure laissez-faire economics.[8]

With the conviction and death of Homer C. Mills in 1958, Searchlight looked to others to bring prosperity to the "camp without a failure." Unfortunately for the town, there was nobody else who could resurrect the gold rush. Prostitution and mining both failed at about the same time. Violation of criminal laws ended mining, and an enactment of law ended prostitution.

Gold production after the war in Searchlight fell dramatically, to 785 ounces in 1946, 737 ounces in 1947, 1,582 ounces in 1948, 1,007 ounces in 1949, 1,620 ounces in 1950; the zenith in the postwar era was 2,041 ounces, and thereafter production fell to practically no output by 1954. The figure for that year provided a compelling contrast with the more than 25,000 ounces produced in 1906.

27

THE WORLD'S OLDEST PROFESSION

Many commercial enterprises rose and fell over the years in Searchlight. Mining had its peaks, but there were more bad days than good. The one business that held steady and that was always present in Searchlight was prostitution. In 1905 Judge R. M. McElwain, of Searchlight, fined a woman for prostitution in an area too close to a school—that is, nearer than four hundred yards, contrary to state law.[1] That same year there was discussion of a citizens committee to draft an ordinance to impose limits on houses of ill fame. In the personal section of the Searchlight paper in 1910, there was a recognition of the death of Mame Brown, who ran a sporting house.[2]

Prostitution was also present in the years between 1910 and 1930.[3] One of Nevada's most famous musicians, Rags M'Goo, whose real name was Harold Gibson, learned to play the piano from a madam named Frankie. In 1977 he was even able to identify the building where Frankie had conducted her trade. Rags bragged that he played professionally his entire life in the style that Frankie taught him.[4]

There was ample evidence of prostitution in the 1920s.[5] And everyone in town knew that prostitution was still going on in the 1930s; the main house was not downtown but stood on the old road to Needles and was operated by Old Man Huff.[6]

The oldest profession achieved new stature in Searchlight just after World War II and continued unabated until sometime in late 1956. During the construction of Davis Dam more than fifty prostitutes were working in Searchlight[7] and thirteen houses of prostitution operated throughout town, with most of the activity concentrated on the Las Vegas–to–Needles Road right in the middle of town. On military payday the cars would fill the streets, and the town would seem to be a hundred times larger than it really was.[8]

Some of the prostitutes worked part-time— that is, they would work weekends as prostitutes and go back to their regular jobs, usually in Las Vegas, during the week.[9]

The houses of prostitution were generally operated most discreetly. A bar or cafe patron could come into the establishment, order a sandwich or drink,and never know of the "girls." The women did not appear in the bars but remained in separate rooms; a client would have to ask for one of the prostitutes, as the women did not solicit and were available only upon request.[10] It was well known that Searchlight was the only place for miles around that had prostitution, except for Roxie's Four Mile on the eastern limits of Las Vegas, which was constantly being raided by the police. Today some people remember nothing about the town of Searchlight but its prostitution, as that was the dominant enterprise during this time.[11]

The biggest and most innovative operator in Searchlight during the height of prostitution was the flamboyant Willie Martello, owner and operator of the El Rey. He came to town in 1947 and became Searchlight's most successful bordello operator. He developed the Searchlight Airport and built the first swimming pool in the town's history. The pool was built for the prostitutes, but Willie reserved it for the kids in Searchlight one day each week, and many Searchlight youths learned to swim there.

Because prostitution was illegal during much of the time that it flourished in Searchlight, there were few written rules regarding how the trade was to be conducted. Rather the peculiar profession was governed by timeworn custom. Not only did the implicit rules in Nevada deal with how a prostitute should behave in public but, more emphatically, custom dictated that these women were to be kept out of sight of the community.

Customs and rules were not always the same throughout the state. In Winnemucca all prostitutes were required to live in one of the licensed brothels except during the one week out of four that they had off. Then they were required to leave town. In this and other communities where prostitution was allowed, the prostitute's family could not live in the town where she worked. Some towns required that prostitutes' cars had to be registered with the local police.[12]

Another unwritten but rigidly observed rule dictated that women who were not prostitutes were forbidden to visit any of the brothels or even drive through the area where the houses stood. This rule was not designed to protect the integrity of the woman; rather, its intent was to prevent wives from driving around the vicinity looking for their husbands. Other rules prevented prostitutes from being accompanied by a male escort when they came to town.[13]

Generally, prostitutes were permitted outside the brothels, with few exceptions, only between certain hours. For example, in Wells the hours were 1 P.M. to 4 P.M. Winnemucca allowed the women in town from 4 A.M. until 7 A.M. There were also restrictions on where they could go once they got to town. They could not frequent bars, residential areas, or gaming houses, nor could they rent rooms in town.[14] Some combination of these unwritten rules was always in effect in the days of Searchlight prostitution.

As Searchlight's preeminent operator, Willie Martello was imaginative, to say the least. Since, even by the 1950s there were still no telephones in town, he solved his communications problem by using carrier pigeons to maintain contact with his family in Las Vegas and to send for supplies and other necessities.[15]

Martello and others who ran houses of prostitution in Searchlight were subject to a rather peculiar Nevada Supreme Court decision with regard to a statute that declared illegal any house of prostitution closer than 400 yards to a school or church or on a main street. The court maintained that the law did not mean that brothels were legal in other parts of town. In short, the court said that such operations could still be declared a public nuisance.[16]

In the 1950s Clark County officials repeatedly tried to close the houses of prostitution in Searchlight, which had been functioning since gold was discovered in 1898. Searchlight fought the closings, and enforcement officials could not get Searchlight residents to file complaints. When the government agents came to town to raid the houses, they found that the owners had been forewarned. On occasion, when a business was ordered closed, it would reopen as soon as the county people left town.[17] The election of Butch Leypolt as sheriff of Clark County and George Dickerson as district attorney, however, led to the end of prostitution in Searchlight in the late 1950s. These two men campaigned on an anti-prostitution platform and followed through with their promise after the election.

No matter how much money Willie Martello made in gambling and prostitution, he was always in debt. One summer day in the mid-1950s, when Marshall Sawyer, the owner of Tranco Products (a company that merged with a New York Stock Exchange company), was flying over the southwestern deserts, he asked, "What is that down there?" The response was that it was Searchlight. "What's there?" The pilot responded, "Whores and gambling."

Thus Sawyer began a venture that had him in Searchlight for the better part of a year. When his spree in Searchlight ended, he had loaned large amounts of money to Willie Martello and had obtained a third deed of trust on Willie's holdings.

Shortly afterward, the El Rey burned to the ground, but only after prostitution had finally ended. The insurance paid off the first and second deeds of trust, but not Sawyer's third. Sawyer foreclosed and graciously made a deal with Martello's friend and successful Las Vegas hotel man, Doc Bailey, to allow Willie to reopen the old Crystal Club, which became the new El Rey. Doc Bailey owned the Hacienda Hotel on the Las Vegas Strip. The new El Rey was smaller and undercapitalized, and so Willie, the generous whoremonger, was forced to abandon his business in the desert. He died shortly thereafter, in Las Vegas in 1968. Martello's last establishment is now the facility used as Searchlight's Senior Citizen Center.

The road from Nipton was finally paved in 1964, after the age of prostitution had ended in Searchlight. The road to Lake Mohave from Searchlight was paved in 1963, allowing travelers from California to visit the great recreational area just below Searchlight with ease. More and more people started leaving overcrowded California to retire in Searchlight. In short, mining ended, as had prostitution, but tourism and retirement slowly began to allow Searchlight to grow once more.

The biggest boon to Searchlight's most recent economic growth was the arrival of Warren and Verlie Doing, who bought a casino in Searchlight. They had been longtime casino operators in other places in Nevada and, seeing the potential of the area, invested in the town in 1967. In effect, they purchased much of the town, and in the late 1970s constructed a million-dollar casino, bar, and restaurant at the intersection of the Las Vegas and Nipton Roads. Warren passed away in 1984, but Verlie put her expertise to work operating the Searchlight Nugget, and with approximately fifty workers has been the largest employer in Searchlight for many years.

In the mid-1960s, on the California–Nevada border south of Searchlight on the Needles Highway, Slim and Nancy Kidwell proved up a section of federal land by drilling wells on the property and striking an underground aquifer. Afterward, they constructed an airfield, mobile home park, bar, cafe, and casino. They were able to get title to the section of land under a since repealed federal law called the Pittman Desert Entry Act. The entire complex was opened in 1968, and today the town they call Cal-Nev-Ari has almost 400 permanent residents.

Searchlight also had a mobile home park with seventy-three spaces, constructed by Junior Cree, who first came to town when he was on his way to a job at Davis Dam. He never made it to the dam, but in 1950 he took up residence in Searchlight, where he has lived for the last forty-five years. Junior's trailer court was opened in the late 1960s.

In recent years, a new elementary school, park, and town hall have been constructed. Through a federal law, the town's very complicated land disputes were finally resolved. The problems arose over the years because people lived on land that they had the right to mine but not to live on. The disputes were turned over to Judge Michael Wendell, a Clark County district judge, who resolved all land boundary problems and created a townsite. Lands left over within the new borders were auctioned and the proceeds spent on the park and town center for Searchlight.

Today, as the town of Searchlight sits on the verge of the twenty-first century, its prospects seem as bright as they were a century ago. It is not known who was the first non-Indian to cross the river below Searchlight; it could have been Ewing Young in 1829, but it could also have been any number of trappers.[18] The first recorded crossing in the area took place in 1858 by Mormon leader Jacob Hamblin, one of Brigham Young's confidants.[19] Hamblin explored the area below Searchlight, and he was noted for his work with Indians throughout the entire Great Basin.[20]

Almost immediately, the area known as Cottonwood Island, a pleasantly lush tract with cottonwood trees and high grasses, located in the middle of the Colorado River, was found to be a good place to raise and pasture animals. Even the U.S. Army assigned a small detachment to tend some of the military livestock. The problem in grazing cattle and horses here was the serious flooding of the river, but despite that, the island was used continuously until it was covered by the lake. In fact, in 1953, it was the site of a mill run by the same people who operated a mine in the Newberry Mountains.[21]

The old Quartette mill, whose foundations were easily distinguished on the shore of the river, was extinguished by the formation of Lake Mohave. Construction on Davis Dam, about twenty miles south of

Searchlight on the Colorado River, got under way in 1942. During the war years the work came almost to a standstill, and most of the construction actually took place between 1946 and 1952. This dam is two hundred feet high, less than one third the size of Hoover Dam. Consequently, its Lake Mohave is also much smaller than Lake Mead.[22]

Before the formation of the lake, the river area was always a recreational site. It was a place for swimming, but most of all, it was used for fishing for catfish, trout, and even turtles. Today the facility is part of the Lake Mead National Recreation Area, offering such activities such as boating, fishing, swimming, and camping. The site has a ranger station, two campgrounds with a total of 145 sites, and a large trailer village of almost 300 sites, with 226 of the pads designated for long-term renters. The marina has 250 boat slips, with additional facilities for renting ski boats, fishing boats, and houseboats. There is a restaurant and a small 24-room lodge. This federal facility has been a boon to the businesses in Searchlight.

The beginning of gambling on the Nevada side of the river, at Davis Dam, has turned out to be very beneficial to Searchlight. Having grown from a few slot machines in a motel to 8,000 hotel rooms, this location now is officially named after its founder, Don Laughlin. This tourist destination began to flourish in the early 1980s, and the traffic level has merited construction of new and improved roads from Las Vegas, allowing many more people to travel to Laughlin, through Searchlight, and thus greatly improving tourism in Searchlight.

Today, Searchlight even has two churches, each in its own permanent building. The Searchlight Community Church, founded in 1965, obtained the land and erected its own building in 1982, a first in the history of the town. At various times Searchlight had had its own church and clergy, but never its own church structure. The other church, the Searchlight Bible Church, was founded in 1989, meeting in various places; it now has its own facility, a building that used to be the post office.

Because of construction costs, most homes are mobile homes or modular homes. In 1993 the federal government, through the Farmers

Home Mortgage Administration (FHMA), now known as the Farm Service Agency, built twenty-seven beautiful homes for qualified senior citizens.

Searchlight has its own water system, with a new well that brings in fresh water from a service well four miles west of town. This project was a joint venture of the Clark County Water District and the federal government. The town has also had a modern sewer system for almost two decades. Willie Martello's airstrip, one mile south of town, has been kept clean and safe and is now used for flights of private aircraft into and out of Searchlight.

The AMVETS Post No. 1 of Nevada meets regularly in Searchlight, as does the Roadrunners Shrine. The VFW has its own post in Searchlight, and a VFW Auxiliary.

Searchlight also has a branch of the Clark County Library. In short, Searchlight is as modern today as it was in 1908. Its businesses include two service stations, a bait store, two small grocery outlets, two beauty shops, two motels, one large trailer park, two smaller ones, and several other small enterprises. For several years, there have been two casinos, but the smaller one recently burned to the ground and is not presently operating.

The most remarkable new addition to the town is the Searchlight Historic Museum. This facility, which is located in the new community center, is a satellite of the Clark County Heritage Museum. The driving force behind the establishment of the museum has been Jane Overy, who with a small grant from the Nevada Humanities Committee and the National Endowment for the Humanities has created a beautiful historical monument to the history of the town.

Law enforcement has changed greatly since the days of Big John Silveria. The town now has its own police substation and justice court. The metropolitan police supply two full-time deputies, who work in conjunction with two full-time Nevada Highway Patrolmen. There are also part-time state and federal game wardens.

Just when it was thought that mining was a thing of the past, a few miles over the Nevada–California border, an old mine has suddenly come back to life. First discovered in 1907 during Searchlight's zenith, the mine was a good enough property to form the town of Hart around the Oro Belle, Jumbo, and Big Chief Mines. In fact, at one time, the Big Chief operated the only mill in the area outside Searchlight. The town reached a population of more than 300 but lasted only four years after the high-grade ore vanished. One account even put the population at between 600 and 700.[23] There was clay in the mines from the beginning and this became a marketable commodity; in the 1920s, the clay was used for popular Franciscan Ware and for the English Wedgwood pottery.[24]

In 1984, a Canadian entrepreneur, Ross Fitzpatrick, formed a company to mine the old property, using modern mining and milling techniques. Construction on the mine began in April 1991, with the first pour of pure gold from its mill occurring in February 1992.[25] The mine has already produced more than 500,000 ounces of gold, and with new discoveries having been made, the life of the project has been forecast to extend well into the next century.

The owner of the mine, the Viceroy Company, has established its Nevada headquarters at the Walking Box Ranch, formerly owned by Rex Bell and Clara Bow. The company has totally restored the property, along with the furniture and the furnishings, to its original 1931 condition. This famous ranch is now listed on the National Register of Historic Places.

28

CONCLUSION

Searchlight never became a ghost town, but it tried. Many have written that a permanent society cannot be built on the industry of mining. Mines become exhausted; ghost towns develop; people move away; societies decay. Cultivating land, tending flocks, developing local industries using local resources—those are the activities that some believe build a permanent community.[1]

The camp that didn't fail had its share of failures. It had a series of booms and busts. It really had only one big boom, though, and that was when the town was discovered.

A mining town is like the ocean. It has tides that sweep in and tides that go out, the ebb and flow of nature. Those in the ebb are sometimes unable to see the flow. But in mining it is always there. There are no permanent towns that survive on mining alone. When the tide goes out, when the boom is over, the debris is all that is left. An old mining camp is just like that. You see old buildings, old gallows frames, old dumps, abandoned roads—the debris left from the boom.

Searchlight was like all mining camps. First we had the discovery of the mineral by the prospector. Then the rush of miners, almost always without women. Next the saloons, then the merchants, including those who sold women to the miners.

The first structures were the tents. Then, if the gold or silver held out, came wooden frame or possibly rock or timber buildings. If the mineral strike was maintained, then came the formation of the town.

The sounds of a mining town are unique. If the hole is being mined near the surface, the sounds of exploding dynamite are clearly discernable. If the mine is deep, the sound is muffled but still distinctive, a kind of low, dull thumping. The old hoists emitted very unusual but clearly recognizable sounds when pulling a bucket or tram of muck or ore to the mine's head. And the stamps of the mills made a distinct low-density thump.

When the town fades, those with money, talent, and initiative generally depart quickly, leaving behind the diehards, the outcasts, the mavericks, or those too old or too sick to move on. Those who remain try to make it on what the big operators left, working the low-grade ore or the good ore that was too dangerous for the previous operators to take.

Mining has changed in Nevada, mainly because of how the minerals are extracted from the ground. Nevada mining operations, until the late 1960s, had virtually all been conducted underground, with the exception of the copper pits at Ely and Yerington. Today the active mines are almost all open pit, using heavy equipment always aboveground, in sight of the sky. However, underground mines are now being developed all over northeastern Nevada. Though some things have changed, the boom-and-bust cycle will prevail; in mining it always does.

Searchlight is now prosperous again, not because of mining but because of the beauty of its landscape, the influx of retirees, and the tourism road, upon which the camp without a failure stands. Not all mining camps in Nevada survived. The names of ghost towns, even in Clark

County, are many, among them Platina, Crescent, Alunite, Callville, West Point, Kaolin, St. Joseph, and Key West—mining camps that ultimately failed and withered away.

Searchlight, like the other towns and cities that started in Nevada because of mining, has a distinctive history, enriched by the people who struggled to make a living in the ground, much of the time in conditions that made mining the most dangerous occupation in America. Searchlight survived it and lives on, a unique camp, a unique town, in a unique place—it didn't fail.

APPENDIX 1

Mineral Production of the Searchlight District, 1902 to 1962

Year	Ore (tons)	Gold (ounces)	Silver (ounces)	Copper (pounds)	Lead (pounds)	Total Value (dollars)
1902	10,910	6,147	1,175	0	0	$127,676
1903	19,522	19,275	27,691	0	0	410,079
1904	16,750	18,400	13,498	0	0	388,068
1905	38,069	19,329	28,528	22,808	12,064	420,931
1906	45,668	25,145	11,543	11,182	9,655	530,227
1907	45,921	23,441	7,494	37,063	38,113	498,947
1908	52,193	13,137	10,883	14,954	44,857	281,199
1909	68,931	16,237	13,544	22,916	36,209	347,224
1910	27,331	10,406	11,489	93,848	45,847	235,237
1911	1,850	966	2,136	12,095	7,659	22,968
1912	4,158	2,237	2,562	10,126	10,032	49,942
1913	5,989	6,654	5,903	53,971	61,006	152,161
1914	3,057	3,999	2,855	5,556	21,919	85,830
1915	7,766	4,985	3,546	20,970	35,562	110,182
1916	6,322	3,903	4,340	24,392	9,951	90,220
1917	12,866	3,445	9,073	98,923	59,453	110,808
1918	1,700	2,649	8,395	44,477	77,863	79,670
1919	6,980	2,854	5,502	18,074	39,307	70,612
1920	2,131	3,432	5,275	44,941	91,436	91,287
1921	1,182	3,601	9,877	31,128	98,048	92,737

1922	1,441	1,661	5,024	12,847	44,375	43,540
1923	1,142	1,007	2,708	22,150	47,651	29,628
1924	850	830	1,267	5,801	46,050	22,453
1925	2,833	3,314	2,586	14,974	86,768	79,975
1926	1,881	1,879	1,720	7,430	81,977	47,522
1927	106	335	306	1,429	10,415	7,940
1928	77	224	252	2,968	9,464	5,573
1929	184	105	58	654	2,743	2,495
1930	124	19	27	0	0	411
1931	20,855	3,276	5,826	6,022	274,687	80,129
1932	2,973	2,018	4,853	4,353	355,458	54,031
1933	2,996	913	1,182	2,261	12,040	19,869
1934	6,701	1,746	8,478	2,237	4,951	67,140
1935	29,617	3,829	12,317	9,195	72,271	146,507
1936	16,705	6,403	36,116	5,580	53,857	255,061
1937	5,692	5,456	28,688	500	12,600	213,954
1938	4,646	4,066	6,871	3,000	9,400	147,478
1939	2,035	1,471	1,682	9,400	14,200	54,272
1940	3,565	3,815	6,462	18,000	14,000	140,854
1941	4,092	3,908	4,524	2,100	5,000	140,530
1942	1,017	897	2,205	0	200	32,976
1943	117	139	1,454	3,000	4,100	6,596
1944	52	21	308	5,200	2,100	1,824
1945	401	852	5,708	2,000	1,200	34,252

Year		Ore (tons)	Gold (ounces)	Silver (ounces)	Copper (pounds)	Lead (pounds)	Total Value (dollars)
1946		393	785	5,937	6,000	4,000	33,680
1947		405	737	2,768	20,300	7,500	33,643
1948		13,436	1,582	1,487	7,800	100	58,427
1949		16,399	1,007	4,109	700	4,500	39,813
1950		18,967	1,620	4,152	1,700	600	60,892
1951		24,850	2,041	5,352	4,800	800	77,579
1952		14,466	667	2,405	5,800	2,400	27,663
1953		2,703	123	379	0	0	4,648
1954		1	4	2	0	0	142
1957	*						
1958	*						
1961	*						
1962	*						
Total		581,018	246,992	352,522	755,625	1,884,388	$6,167,502

*some small production reported but the actual amounts may not be disclosed.
Source: Mineral Resources of the United States and Mineral Yearbook.

APPENDIX 2

Appendix 2.1. Total Precipitation (inches) for Searchlight, 1949–1993

Year	Jan	Feb	Mar	Apr	May	Jun	Jul	Aug	Sep	Oct	Nov	Dec	Total
1949	3.67	0.94	0.35	0.13	0.18	0.20	0.63	0.70	0.00	0.05	0.32	0.55	7.72
1950	0.14	0.05	0.80	0.00	0.00	0.00	0.57	0.03	0.14	0.00	0.06	0.04	1.83
1951	0.78	0.00	0.08	0.65	0.93	0.00	2.23	1.04	1.49	0.52	0.57	1.32	9.61
1952	1.59	0.00	3.09	1.90	0.00	0.22	0.50	0.26	2.68	0.00	0.67	0.88	11.79
1953	0.02	0.28	0.09	0.13	0.00	0.00	1.37	1.01	0.00	0.10	0.00	0.09	3.09
1954	1.55	0.25	2.81	0.00	0.00	0.00	2.10	2.54	0.22	0.00	0.40	0.45	10.32
1955	2.35	0.17	0.00	0.05	0.04	0.00	1.74	2.78	0.00	0.00	0.12	0.15	7.40
1956	0.39	0.00	0.00	0.12	0.00	0.00	0.28	0.04	0.04	0.77	0.00	0.00	1.64
1957	0.63	0.40	0.13	1.85	0.28	0.45	0.62	1.49	0.00	1.90	0.92	0.30	8.97
1958	0.33	1.48	1.97	0.79	1.22	0.27	0.60	0.05	1.81	0.70	0.56	0.00	9.78
1959	0.19	1.45	0.00	0.00	0.10	0.00	1.86	0.50	0.42	1.33	1.09	2.55	9.49
1960	1.39	0.92	0.00	0.05	0.00	0.00	0.06	0.00	0.29	0.38	3.15	0.50	6.74
1961	0.41	0.00	0.07	0.14	0.00	0.00	0.13	1.15	0.30	0.53	0.05	0.13	2.91
1962	0.04	0.84	0.05	0.00	0.42	0.44	0.05	0.43	0.37	0.26	0.02	0.29	3.21
1963	0.21	0.84	0.26	0.30	0.00	0.06	0.00	0.77	1.29	1.27	0.30	0.00	5.30
1964	0.00	0.00	0.51	0.44	0.04	0.33	0.32	0.00	0.00	0.00	0.31	0.00	1.95
1965	0.00	0.70	1.09	3.89	0.14	0.00	0.66	1.95	0.00	0.00	1.96	2.57	12.96
1966	0.10	0.49	0.57	0.00	0.00	0.08	0.75	0.47	0.44	0.46	0.17	1.96	5.49
1967	0.64	0.00	0.00	0.19	0.10	0.00	0.69	3.12	0.11	0.00	0.94	1.07	6.86
1968	0.00	0.39	0.50	0.00	0.04	0.30	1.78	0.16	0.00	0.00	0.28	0.03	3.48
1969	2.77	1.36	0.49	0.00	0.92	0.50	0.83	0.10	1.15	0.12	0.26	0.00	8.50
1970	0.05	0.51	1.00	0.00	0.00	0.08	0.25	2.29	0.00	0.00	0.28	0.57	5.03

1971	0.00	0.36	0.00	0.00	1.01	0.00	0.00	0.67	0.00	0.06	0.06	0.55	2.71
1972	M	0.00	0.00	0.53	0.15	0.40	0.00	0.94	0.40	1.01	1.54	0.41	5.38
1973	0.53	1.79	1.47	0.00	0.24	0.40	0.00	0.09	0.00	0.00	0.32	M	4.84
1974	1.48	0.05	0.64	0.02	0.00	0.00	2.45	0.16	0.10	1.32	0.33	0.71	7.26
1975	0.05	0.11	1.42	0.15	0.15	0.00	0.00	1.19	2.13	0.45	0.00	0.00	5.65
1976	0.00	2.29	0.48	0.67	0.90	0.00	1.00	0.00	5.93	1.00	0.03	0.00	12.30
1977	0.47	0.00	0.57	0.06	0.46	0.04	0.69	1.13	0.00	0.19	0.00	0.63	4.24
1978	2.15	1.92	2.35	0.80	0.00	0.00	1.21	4.11	0.69	0.94	2.81	1.67	18.65
1979	1.24	0.47	2.69	0.00	0.03	0.10	3.23	0.94	0.00	0.50	0.00	0.90	10.10
1980	2.71	4.95	1.15	1.34	0.25	0.00	1.06	0.02	0.73	0.13	0.00	0.00	12.34
1981	0.08	0.37	2.06	0.00	0.90	0.09	0.45	0.65	0.94	0.32	0.53	0.02	6.41
1982	0.50	0.82	1.79	0.24	0.28	0.00	0.17	7.82	0.27	0.52	1.09	0.98	14.48
1983	2.54	0.77	2.61	0.92	0.03	0.00	1.00	3.59	0.95	1.43	0.01	0.98	14.83
1984	0.00	0.00	0.00	0.06	0.00	0.17	4.97	0.66	0.18	0.12	1.60	4.07	11.83
1985	1.44	0.28	0.03	0.80	0.03	0.04	1.08	0.00	0.65	0.40	1.36	0.07	6.18
1986	0.52	0.55	0.71	0.03	0.00	0.00	0.17	0.73	0.62	0.56	0.27	3.00	7.16
1987	0.67	0.82	0.24	0.19	0.32	0.04	0.71	0.00	0.32	2.25	1.43	0.84	7.83
1988	0.88	0.58	0.00	1.53	0.03	0.09	0.54	1.45	0.00	0.00	0.39	0.16	5.65
1989	1.27	0.28	0.12	0.00	0.14	0.00	0.83	0.24	0.00	0.08	0.00	0.06	3.02
1990	1.17	0.21	0.13	0.10	0.24	0.48	1.66	1.38	0.47	0.11	0.04	0.00	5.99
1991	1.29	1.50	2.11	0.00	0.35	0.01	0.13	0.71	0.67	0.19	0.23	0.84	8.03
1992	1.23	2.56	5.07	0.04	0.52	0.05	0.09	1.20	0.00	2.10	0.00	1.62	14.48
1993	3.72	3.86	0.66	0.00	0.10	0.17	0.00	2.34	0.00	0.05	0.46	0.07	11.43

M = missing data
Source: Courtesy of John W. James, Nevada State Climatologist.

Appendix 2.2. Average Temperature (Fahrenheit) for Searchlight, 1949–1993

Year	Jan	Feb	Mar	Apr	May	Jun	Jul	Aug	Sep	Oct	Nov	Dec	Mean
1949	31.6	40.7	51.8	65.3	68.0	79.1	84.9	82.7	79.8	63.7	62.1	45.4	62.9
1950	41.5	52.5	56.1	67.0	70.4	77.9	83.2	83.2	74.9	70.5	58.3	53.1	65.7
1951	44.5	46.9	52.1	61.1	69.1	77.2	85.0	80.8	78.6	64.0	50.8	40.9	62.6
1952	40.5	46.0	45.9	59.5	72.0	74.0	83.1	84.3	77.9	71.6	48.5	44.1	62.3
1953	49.8	47.8	54.6	59.6	61.8	76.2	84.9	81.5	79.8	64.7	56.4	44.9	63.5
1954	45.4	54.7	50.3	65.6	71.9	76.2	85.0	80.1	77.0	67.2	56.9	44.7	64.6
1955	39.4	43.2	52.1	57.1	66.7	77.0	80.7	80.3	77.5	68.9	53.4	48.3	62.0
1956	48.6	43.5	56.1	59.8	69.8	80.2	82.6	81.0	80.8	63.6	52.0	47.0	63.8
1957	41.3	52.4	54.5	59.0	63.3	80.7	85.5	81.2	77.5	60.7	47.8	48.2	62.7
1958	46.1	49.9	45.9	58.0	73.1	78.1	83.1	84.9	77.3	69.0	53.0	50.3	64.1
1959	47.4	44.7	55.5	64.6	67.7	81.7	86.8	80.7	73.9	66.5	54.2	46.3	64.2
1960	38.8	42.6	56.4	63.6	69.5	82.9	86.6	84.1	78.7	64.8	53.1	45.5	63.9
1961	47.6	51.0	54.0	62.5	67.5	81.7	86.3	82.9	73.1	66.3	48.6	42.2	63.6
1962	43.7	47.8	48.1	67.2	66.3	76.6	82.1	83.8	76.9	67.0	58.1	48.0	63.8
1963	42.9	56.0	51.1	57.8	71.9	75.6	84.8	82.9	77.3	67.8	51.9	45.6	63.8
1964	40.6	43.1	47.6	58.0	65.8	76.7	85.5	85.5	M	71.8	47.8	45.4	60.7
1965	46.9	48.3	49.9	55.5	65.1	72.0	82.7	81.9	70.8	69.8	54.4	44.0	61.8
1966	40.2	43.3	54.0	62.7	72.1	78.6	83.7	84.4	76.5	65.3	54.3	44.5	63.3
1967	45.5	50.0	54.8	50.8	68.3	74.1	86.1	83.5	75.7	68.5	56.6	39.3	62.8
1968	45.1	53.2	54.6	58.6	68.6	79.3	82.3	78.2	76.5	66.4	53.4	41.7	63.2
1969	47.2	43.8	49.7	60.4	70.9	75.3	84.0	86.5	77.7	61.3	53.5	47.9	63.2

1970	45.0	50.9	51.8	55.4	70.6	79.4	86.2	83.6	74.2	62.5	53.9	43.3	63.1
1971	45.0	48.6	54.0	58.5	63.4	76.6	86.6	81.7	74.8	59.9	48.7	39.0	61.4
1972	M	51.1	62.4	61.3	69.5	80.4	87.2	82.1	73.7	59.5	46.7	41.1	65.0
1973	40.6	45.7	46.1	58.2	72.8	80.6	85.9	83.3	75.8	66.0	52.0	47.5	62.9
1974	40.6	47.4	55.1	60.6	71.7	84.8	84.1	83.0	78.1	64.9	52.3	42.1	63.7
1975	44.9	46.3	49.6	54.7	68.3	77.9	84.5	83.7	78.0	64.0	53.4	47.9	62.8
1976	47.5	50.9	53.4	57.7	71.8	78.1	83.9	79.8	71.1	61.2	56.1	45.7	63.1
1977	45.5	52.8	48.5	63.9	62.8	82.1	85.5	84.2	76.2	69.0	56.8	49.6	64.7
1978	45.2	48.2	54.8	58.0	68.7	82.1	85.8	82.4	74.8	70.2	49.8	41.0	63.4
1979	37.7	45.2	51.5	59.8	68.5	80.5	84.8	80.1	77.6	68.0	50.4	49.0	62.8
1980	45.6	49.7	49.3	60.5	64.4	77.6	86.1	82.9	77.0	66.4	56.5	53.2	64.1
1981	49.4	51.8	52.5	66.0	69.9	84.4	85.3	85.0	77.3	62.8	57.5	49.6	66.0
1982	42.1	48.8	50.7	58.3	68.3	76.7	82.5	82.1	73.2	61.4	49.3	42.3	61.3
1983	47.5	48.1	51.4	54.0	69.2	76.2	82.5	78.9	77.4	64.6	51.9	46.7	62.4
1984	47.0	49.6	55.2	59.5	75.5	77.3	82.2	82.4	79.3	59.0	51.0	41.9	63.3
1985	42.7	46.3	50.4	65.0	70.4	81.8	85.3	82.0	70.3	63.2	47.3	45.6	62.5
1986	51.2	51.5	58.4	61.2	71.9	82.3	81.8	83.2	67.9	60.9	53.1	45.7	64.1
1987	43.7	47.4	51.4	66.3	69.2	81.1	80.2	81.8	77.2	68.6	51.4	40.2	63.2
1988	44.0	50.5	54.5	60.0	69.7	78.8	85.6	81.4	75.7	72.3	53.4	44.5	64.2
1989	42.4	47.2	59.1	68.9	71.0	80.1	87.4	81.2	76.4	65.0	55.9	48.7	65.3
1990	44.6	46.0	56.5	63.5	68.3	81.1	84.7	80.9	78.0	66.8	53.9	40.0	63.7
1991	44.5	55.5	47.5	58.9	65.4	74.8	84.1	82.6	76.8	69.0	53.5	44.6	63.1
1992	44.7	50.4	51.9	64.0	71.0	76.7	82.3	83.8	78.0	67.8	49.5	29.7	62.5
1993	43.2	45.7	55.6	62.4	70.1	77.5	80.4	80.8	76.6	61.5	50.3	44.0	62.3

M = missing data
Source: Courtesy of John W. James, Nevada State Climatologist.

Appendix 2.3. Searchlight Climatological Summary

MEANS: 1961 - 1990
EXTREMES: 1915 - 1993

LATITUDE 35° 28' N
LONGITUDE 114° 55' W
ELEVATION 3540 FEET

	TEMPERATURE							MEAN DEGREE DAYS**	PRECIPITATION TOTALS (INCHES)								MEAN NUMBER OF DAYS					
	MEANS			EXTREMES								SNOW & SLEET						TEMPERATURES				
																		MAXIMUM		MINIMUM		
	DAILY MAXIMUM	DAILY MINIMUM	MONTHLY	RECORD HIGHEST	YEAR	RECORD LOWEST	YEAR		MEAN	GREATEST DAILY	YEAR	MEAN	MAXIMUM MONTHLY	YEAR	GREATEST DAILY	YEAR	PRECIPITATION .10" OR MORE	90° AND ABOVE	32° AND BELOW	32° AND BELOW	0° AND BELOW	MONTH
JAN	53.8	35.1	44.5	77	1971	7	1937	636	.94	1.29	1955	1.8	38.0	1949	14.0	1949	2	0	0	8	0	JAN
FEB	59.2	39.2	48.7	81	1986	6	1933	456	.81	1.97	1980	0.0	2.0	1949+	2.0	1949	3	0	0	5	0	FEB
MAR	64.6	40.8	52.7	90	1981	20	1966	392	.91	1.74	1954	0.2	3.5	1977	3.5	1977	4	0	0	3	0	MAR
APR	73.2	47.1	60.2	94	1982	27	1945	210	.41	1.32	1957	0.0	1.0	1967	1.0	1967	0	0	0	0	0	APR
MAY	82.7	55.4	69.1	102	1974	30	1915	41	.24	1.17	1958	0.0	0.0	1989	0.0	1993	1	2	0	0	0	MAY
JUN	93.0	65.0	79.0	110	1924	40	1932	0	.11	0.48	1969	0.0	0.0	1989	0.0	1993	0	17	0	0	0	JUN
JUL	97.9	71.1	84.5	111	1942	52	1987	0	.90	2.07	1979	0.0	0.0	1989	0.0	1993	1	28	0	0	0	JUL
AUG	95.9	69.5	82.5	110	1933	51	1920	0	1.17	4.50	1982	0.0	0.0	1989	0.0	1993	2	24	0	0	0	AUG
SEP	88.3	62.5	75.4	107	1950	41	1986	0	.56	3.89+	1939	0.0	0.0	1976	0.0	1993	1	15	0	0	0	SEP
OCT	77.3	53.4	65.3	98	1978	23	1971	115	.50	1.53	1992	0.0	0.0	1989	0.0	1993	1	5	0	0	0	OCT
NOV	63.2	42.1	52.6	86	1924	15	1919	378	.57	2.25	1960	0.2	6.0	1964	4.0	1964	1	0	0	1	0	NOV
DEC	53.9	35.6	44.8	78	1928	8	1990	626	.71	2.44	1986	0.5	6.0	1967	5.0	1960	2	0	0	11	0	DEC
YEAR	75.2	51.3	63.3	111	1942	6	1933	2854	7.84	4.50	1982	2.5	38.0	1949	14.0	1949	18	91	0	30	0	YEAR

T Trace, an amount too small to measure
** Base 65°

\+ Also on earlier dates, months, or years
* Less than one-half

APPENDIX 3

Postmasters, Searchlight, Clark County, Nevada (since 1926)

Name	Title	Date Appointed
Light Wheatley	Acting Postmaster	October 2, 1926
Light Wheatley	Postmaster	December 13, 1926
Robert L. Fenlon	Acting Postmaster	December 13, 1930
Miss Bernice M. Kamer*	Postmaster	April 11, 1931
Miss Reen Lee	Acting Postmaster	October 31, 1945
Mrs. Wilberta F. Silveira	Acting Postmaster	May 25, 1946
Mrs. Wilberta F. Silveira	Postmaster	February 4, 1947
Douglas M. Hoopes	Acting Postmaster	June 29, 1948
Douglas M. Hoopes	Postmaster	May 5, 1949
Mrs. Wilberta G. Reid	Acting Postmaster	July 28, 1951
Mrs. Wilberta G. Reid	Postmaster	March 28, 1952
Ms. Mary C. Valenti	Acting Postmaster	December 6, 1960
Mrs. Wilberta G. Reid	Acting Postmaster	May 18, 1962
Mrs. Wilberta G. Reid	Postmaster	July 16, 1962
Mrs. Dorothy H. McCoy	Acting Postmaster	November 1, 1967
Mrs. Dorothy H. McCoy	Postmaster	December 5, 1970
Frederick J. Vogt	Officer-in-Charge	October 17, 1980
Helen Johnson	Postmaster	November 15, 1980
Erma L. Antrim	Officer-in-Charge	April 6, 1984
Erma L. Antrim	Postmaster	December 8, 1984
Mrs. Shelby Jean Auflere	Officer-in-Charge	April 1, 1988
Mrs. Marlys A. Thompson	Officer-in-Charge	April 11, 1988
Mrs. Shelby Jean Auflere	Postmaster	July 30, 1988

*Miss Bernice M. Kamer's name was changed to Bernice M. Fenlon by marriage on November 6, 1935.

Source: USPS Historian, Corporate Information. Courtesy of the U.S. Postal Service.

NOTES

1: THE BEGINNING

1. Townley, "Early Development," 9.

2. Bancroft, *History of Arizona,* 482–502.

3. Price, *War of the Rebellion,* 355–60.

4. Casebier, *Guide to the East Mojave Heritage Trail,* 223.

5. Price, *War of the Rebellion,* 357.

6. Casebier, *Guide to the East Mojave Heritage Trail,* 223.

7. Ibid., 25.

8. Townley, "Early Development," 9–23.

9. Riggs, *Reign of Violence in Eldorado Canyon,* 98.

10. Scrugham, *History of Nevada,* 611.

11. Articles of Incorporation, Piute Company, June 30, 1870.

12. Lingenfelter, *Steamboats of the Colorado,* 68–70.

2: MONEY FROM MASSACHUSETTS

1. Townley, "Early Development," 21.

2. Ibid., 22.

3. Mary Ellen Sadovich, "Searchlight," *Las Vegas Review-Journal—The Nevadan,* 30 April 1967, 24.

4. *Searchlight,* 13 August 1905; *Bulletin,* 8 February 1907.

5. *Bulletin,* 8 February 1907.

6. Ibid.

7. *Las Vegas Review-Journal,* 30 April 1967, 24.

8. *Searchlight,* 13 August 1905.

9. Elliott, *Nevada's Twentieth-Century Mining Boom,* 183.

3: THE CAMP THAT DIDN'T FAIL

1. *Engineering and Mining Journal,* 2 July 1898, 18.
2. Ibid. See also Townley, "Early Development," 23.
3. *Searchlight,* 29 July 1906.
4. *Bulletin,* 19 July 1907.
5. Ibid.
6. Ibid.
7. Ibid., 14 July 1911.
8. Ibid.
9. Paher, *Nevada Ghost Towns and Mining,* 280.
10. Townley, "Early Development," 22–23.
11. *Engineering and Mining Journal,* 17 November 1900, 590; see also Townley, "Early Development," 23.
12. *Bulletin,* 8 February 1907.
13. *Searchlight,* 13 August 1905.
14. Townley, "Early Development," 22; *Bulletin,* 8 February 1907.
15. Paher, *Nevada Ghost Towns and Mining Camps,* 282.
16. *Searchlight,* 13 August 1905.
17. Charles Jonas, letter to Mollin Investment Company, 9 November 1934.

4: WHY SEARCHLIGHT?

1. *Engineering and Mining Journal,* 11 February 1899, 184; see also Townley, "Early Development," 23.
2. *Bulletin,* 7 December 1906.
3. Mary Ellen Sadovich, "Searchlight," *Las Vegas Review-Journal—The Nevadan,* 30 April 1967, 24.
4. *Searchlight,* 5 January 1906.
5. *Bulletin,* 14 July 1907.
6. Townley, "Early Development," 22.
7. *Bulletin,* 14 July 1911.
8. Townley, "Early Development," 22.
9. *Searchlight,* 5 October 1906.
10. Ibid.
11. *Bulletin,* 7 June 1912.

5: THE BIG STRIKE

1. Elliott, *Nevada's Twentieth-Century Mining Boom,* 103–48.
2. *DeLamar Lode,* 16 June 1903.
3. *Searchlight,* 26 June 1903.
4. Ibid.
5. Ibid.
6. Ibid., 10 July 1903.
7. Ibid., 28 August 1903.
8. Ibid., 2 October 1903, and *DeLamar Lode,* 17 November 1903.
9. Elliott, *Nevada's Twentieth-Century Mining Boom,* 239.
10. L. W. "Joe" Lappin, letter to author, 1 November 1994.
11. *Searchlight,* 18 December 1903.

6: THE BIG MINE

1. *Searchlight,* 27 May 1904.
2. Ibid.
3. *Las Vegas Sun,* 1 March 1970; see also Townley, "Early Development"; *Bulletin,* 8 February 1907.
4. *Bulletin,* 8 February 1907.
5. Ibid., 13 and 27 December 1907.
6. Ibid., 8 February 1907.
7. *Searchlight,* 5 February 1904.
8. Ibid., 23 September 1904.
9. Ibid., 24 June 1904.
10. Ibid., 4 March 1904.
11. Ibid. 2 February 1906.
12. Ibid., 24 June 1904.
13. Ibid., 15 September 1905.
14. Ibid., 3 February 1905.
15. Ibid., 25 August 1905.
16. Lord, *Comstock Mining and Miners,* 90.
17. *Bulletin,* 9 November 1906.
18. *Searchlight,* 16 June 1905.
19. Ibid., 15 December 1905.
20. Ibid., 25 August 1905.
21. *Bulletin,* 28 February 1908.
22. *Searchlight,* 3 November 1905.

23. *Bulletin,* 6 March 1908.

24. *Searchlight,* 26 October 1906; *Bulletin,* 21 December 1906.

25. *Bulletin,* 6 August 1909.

26. *Searchlight,* 24 August 1906.

27. Eugene Callaghan, "Geology of the Searchlight District, Clark County, Nevada," *Geological Survey Bulletin 906-D* (1939): 135–88.

28. Charles Jonas, letter to Mollin Investment Company, 9 November 1934.

29. *Bulletin,* 31 December 1909.

30. Ibid., 4 December 1908.

31. Ibid., 21 January 1910, 8 February 1910, 10 June 1910.

32. Ibid., 5 July 1912, 4 October 1912.

33. Ibid., 19 November 1911.

34. Ibid., 5, 12, and 19 January 1912, 16 February 1912.

35. Elliott, *Nevada's Twentieth-Century Mining Boom,* 13–15.

36. Jonas to Mollin.

37. Terry Hudgens, interview with Don Reid, 13 March 1996.

38. Jonas to Mollin.

7: THE COMING OF THE RAILROAD

1. *Bulletin,* 7 June 1912.

2. Casebier, *Goffs and Its Schoolhouse,* 31–33.

3. *Searchlight,* 13 November 1903.

4. Myrick, *Railroads of Nevada and Eastern California,* 2:848–49.

5. Ibid., 849.

6. *Lahontan Valley News,* 10 November 1987, 10.

7. *Bulletin,* 10 March 1910.

8. *Searchlight,* 10 June 1904, 16 December 1904, 10 and 31 March 1905.

9. The bulk of this information was obtained from the records of the Santa Fe Company, particularly from E. P. Ripley, the president of the western operations of the company, and Victor Morawetz, the president of the parent company, who was based in New York City. Most of these communications were conducted by telegraph. This correspondence was obtained from Dennis Casebier, who received the material from Bert Foreman in 1970. Foreman was at one time the CEO of the Arizona Historical Society at Arizona State University. The originals of these records are in the Santa Fe collection of the Kansas State Historical Society in Topeka.

10. E. P. Ripley, letter to Victor Morawetz, 9 March 1905.

11. Ibid., 29 March 1905.

12. E. P. Ripley, telegram to Victor Morawetz, 9 March 1906.

13. *Bulletin,* 2 November 1906.

14. Ibid.,18 January 1907.

15. Myrick, *Railroads of Nevada and Eastern California,* 2:851.

16. Haenszel, *Searchlight Remembered,* 22.

17. *Bulletin,* 5 April 1907.

18. *Lahontan Valley News,* 10 November 1987.

19. *Bulletin,* 25 December 1908.

20. Townley, "Early Development," 25.

8: A FASHIONABLE TOWN

1. *Bulletin,* 8 February 1907.

2. Ibid., 6 June 1912.

3. Jane Ann Morrison, "Spotlight on Searchlight," *Las Vegas Review-Journal—The Nevadan,* 9 October 1977, 3J.

4. *Las Vegas Sun,* 6 July 1980, 32.

5. *Searchlight,* 14 April 1905.

6. Ibid., 15 September 1905.

7. Ibid., 23 June 1905.

8. *Bulletin,* 17 December 1909.

9. *Searchlight,* 13 November 1903.

10. Ibid., 4 November 1905.

11. Ibid.

12. Ibid., 22 July 1906.

13. Ibid., 24 November 1905.

14. *Bulletin,* 5 February 1909.

15. Ibid.

16. Elliott, *Nevada's Twentieth-Century Mining Boom,* 56, 240.

17. "Searchlight," *Las Vegas Sun Magazine,* 20 December 1981, 5.

18. *Searchlight,* 24 February 1905, 10 March 1905, 23 April 1905.

19. Ibid., 18 March 1904.

20. Ibid., 16 March 1906.

21. Ibid., 23 March 1906.

22. *Bulletin,* 6 September 1907.

23. Ibid., 1 January 1909.

24. Ibid., 17 December 1909.

25. *Searchlight,* 4, 11, and 18 March 1904.

26. Ibid., 8 July 1904.

27. *Bulletin,* 22 October 1909.

28. *Searchlight,* 1 July 1904.

29. Elliott, *Nevada's Twentieth-Century Mining Boom,* 50, 245.

30. *Bulletin,* 3 January 1908.

31. Jane Overy, interview with author, 5 May 1996.

9: NEWSPAPERS

1. *Searchlight,* 26 October 1906.

2. *Bulletin,* 12 June 1912.

3. Sally Macready Wallace, interview with author, 15 July 1995.

4. *Bulletin,* 4 January 1907.

5. Lingenfelter and Gash, *The Newspapers of Nevada,* 225.

6. Elliott, *Nevada's Twentieth-Century Mining Boom,* 79–80.

7. Ibid., 276.

10: EVEN A BANK

1. Moody, *Masters of Capital,* 142.

2. *Searchlight,* 16 June 1905.

3. Ibid., 28 July 1905, 13 August 1905.

4. Ibid., 8 December 1905.

5. Ibid., 23 March 1906.

6. Ibid., 25 May 1906.

7. *Bulletin,* 1 November 1907.

8. Ostrander, *Nevada,* 137–38.

9. *Bulletin,* 13 December 1907.

10. Ibid., 27 December 1907.

11. *Searchlight,* 7 July 1905.

12. *Bulletin,* 20 November 1908.

13. Ibid., 29 January 1909, 23 April 1909.

14. Ibid., 31 January 1908.

15. Ibid., 1 April 1910.

11: BOOMED OUT

1. "Nevada in the Making," *Nevada State Historical Society Papers,* 1923–24, 377–78.

2. Ibid., 378.

3. Jane Ann Morrison, "Recollections of a Boyhood in Searchlight," *Las Vegas Review-Journal—The Nevadan,* 25 December 1977.

4. *Las Vegas Sun,* 7 September 1975, 3.

5. "Vegas Vignettes," *Las Vegas Review-Journal,* 4 December 1947.

6. *Bulletin,* 6 December 1912.

7. Bob Kerwin, "Reminiscences of Searchlight, 1908–1915," *Nevada Historical Quarterly* (Spring 1976).

8. Ibid.

12: OTHER CAMPS

1. Carlson, *Nevada Place Names,* 221, 231.

2. Ibid., 90.

3. Jane Ann Morrison, "Recollections of a Boyhood in Searchlight," *Las Vegas Review-Journal—The Nevadan,* 25 December 1977, 4J.

4. Casebier, *Guide to the East Mojave Heritage Trail,* 279.

5. *Bulletin,* 13 and 27 December 1907.

13: CATTLE AND CROPS

1. Casebier, *Guide to the East Mojave Heritage Trail,* 129.

2. Bishop, *The Castle Mountain Story,* 69.

3. Casebier, *Guide to the East Mojave Heritage Trail,* 129.

4. *Bulletin,* 22 March 1907.

5. Dennis Casebier, letter to author, 3 May 1996.

6. *Bulletin,* 22 March 1907.

7. Arda Haenszel, *Searchlight Remembered,* 6.

8. Casebier, letter to author, 3 May 1996.

9. *Searchlight,* 8 September 1905.

10. Casebier, letter to author, 3 May 1996.

11. *Bulletin,* 25 March 1910.

12. Hulse, *The Nevada Adventure,* 157.

13. *Bulletin,* 17 November 1911.

14. Ibid.

15. Bishop, *The Castle Mountain Story,* 73–75.

16. Haenszel, *Searchlight Remembered,* 31.

17. *Searchlight,* 8 September 1905.

18. Ibid., 15 September 1905, 8 September 1906.

19. *Bulletin,* 23 July 1909.

20. Velma Verzanni Horne, interview with author, 4 September 1995.

21. *Bulletin,* 18 November 1910.

22. *Searchlight,* 29 September 1905.

23. *Bulletin,* 17 July 1911.
24. *Searchlight,* 5 April 1905.
25. *Bulletin,* 4 August 1911.
26. Ibid., 11 January 1907.

14: COMMERCIAL COMPETITION

1. *Bulletin,* 23 November 1906.
2. Ibid., 18 January 1907.
3. Ibid., 26 February 1909.
4. *Searchlight,* 11 May 1906.
5. Ibid., 26 June 1903.
6. Ibid., 11 June 1905.
7. *Ely Daily Expositor,* 12 January 1909.
8. *Searchlight,* 3 November 1905.
9. Ibid., 24 November 1905.
10. Ibid., 8 December 1905.
11. Ibid., 9 February 1906.
12. Ibid., 8 June 1906.
13. Ibid., 23 March 1906.
14. Ibid., 27 April 1906.
15. Ibid., 22 and 29 June 1906.
16. Ibid., 14 September 1906.
17. *Bulletin,* 16 November 1906.
18. Ibid., 23 November 1906.
19. *Searchlight,* 22 June 1906.
20. *Bulletin,* 23 November 1906.
21. *Searchlight,* 26 June 1903.
22. Ibid., 3 June 1904.
23. Elliott, *Nevada's Twentieth-Century Mining Boom,* 186.
24. *Searchlight,* 2 September 1904.
25. Ibid., 3 June 1904, 2 September 1904.
26. Ibid., 13 November 1903.
27. Ibid., 27 January 1905, 3 February 1905.
28. Ibid., 2 November 1906.
29. *Bulletin,* 22 March 1907.
30. Ibid., 2 November 1906.
31. Ibid., 17 January 1908.
32. *Searchlight,* 28 April 1905.

15: POLITICS

1. Henry Hudson Lee, "Memoirs by Henry Hudson Lee," paper presented at Nevada Historical Society Meeting, 24 January 1968, Las Vegas.

2. *Searchlight,* 5 October 1906; *Bulletin,* 21 October 1910.

3. *Searchlight,* 20 May 1904.

4. Ibid., 12 October 1906.

5. Ibid., 5 October 1906.

6. Ibid., 7 September 1906.

7. Ibid., 22 July 1904, 15 June 1906, 2 August 1906.

8. Ibid., 17 and 24 August 1906, 12 October 1906.

9. *Bulletin,* 19 March 1909.

10. Ibid.

11. Ibid., 2 August 1907, 11 October 1907, 22 November 1907.

12. *Las Vegas Voice,* 4 August 1939, 1.

16: SCHOOLS

1. Elliott, *Nevada's Twentieth-Century Mining Boom,* 49–50.

2. *Searchlight,* 20 May 1904.

3. Ibid., 7 September 1906.

4. *Bulletin,* 25 January 1907.

5. "Section 1350—An Act to Compel Children to Attend School," in *Compiled Laws of Nevada, 1861–1900,* compiled and annotated by Henry Cutting (Carson City, Nev.: Superintendent of State Printing, 1900), 1380.

6. *Bulletin,* 13 September 1912.

7. Brian Cram, Superintendent of the Clark County School District, interview with author, 10 September 1995.

8. Haenszel, *Searchlight Remembered,* 60–64.

9. Jeff Reid, interview with Eleanor Johnson, 10 October 1980.

10. Harvey Dondero, *History of Clark County Schools,* ed. Billie F. Shank (Las Vegas: Clark County School District, 1986) 36–37.

11. Orvis Ring, Superintendent of Public Instruction, *State of Nevada Biennial Report of the Superintendent of Public Instruction* (Carson City: State Printing Office, 1903), 14.

12. Orvis Ring, Superintendent of Public Instruction, *State of Nevada Biennial Report of the Superintendent of Public Instruction* (Carson City: State Printing Office, 1905), 7.

13. John Edwards Bray, Superintendent of Public Instruction, *State of Nevada Biennial Report of the Superintendent of Public Instruction* (Carson City: State Printing Office, 1913), 52.

14. John Edwards Bray, Superintendent of Public Instruction, *State of Nevada Biennial Report of the Superintendent of Public Instruction* (Carson City: State Printing Office, 1915), 51.

17: CRIME AND PUNISHMENT, OR THE LACK THEREOF

1. *Searchlight,* 20 November 1903, 18 December 1903.

2. Bancroft, *History of Nevada,* 29.

3. *Searchlight,* 14 August 1903.

4. Ibid., 27 May 1904.

5. *Bulletin,* 26 June 1908.

6. *Searchlight,* 23 October 1903.

7. Elliott, *Nevada's Twentieth-Century Mining Boom,* 116; see also Joseph R. Conlin, "Goldfield and High-Grade," *American West,* May/June 1983, 38–44.

8. Elliott, *Nevada's Twentieth-Century Mining Boom,* 115.

9. *Searchlight,* 29 June 1906, 17 February 1905.

10. *Bulletin,* 10 January 1908, 12 January 1912; *Searchlight,* 17 February 1905.

11. *Bulletin,* 25 October 1907, 8 November 1907, 6 December 1907.

12. Ibid., 22 January 1909.

13. Carl Myers, interview with Don Reid, 22 March 1995.

14. Erlene Smith, interview with author, 28 May 1995.

15. Terry Hudgens, interview with Don Reid, 9 March 1996.

18: ACCIDENTS

1. *Bulletin,* 28 December 1906.

2. Ibid., 23 July 1909, 1 April 1910, 30 August 1912.

3. Ibid., 12 and 18 November 1910.

4. Ibid., 30 August 1912.

5. "The Coal Mine Operator and Safety," *Labor History* 14, no. 4 (Fall 1973): 483–505.

6. Ella S. Kay, interview with Dennis Casebier, 13 September 1912.

7. Terry Hudgens, interview with Don Reid, 13 March 1996.

8. *Bulletin,* 3 July 1908.

9. Ibid., 1 December 1910.

10. Ibid., 3 April 1911, 19 July 1912.

19: WEATHER

1. Office of Nevada State Climatologist, July 1994.

2. *Searchlight,* 24 August 1906.

3. *Bulletin,* 7 January 1910.

4. Ibid., 3 June 1910.

5. Ibid., 21 July 1911.

6. John W. James, Climatological Summary for Searchlight, Office of Nevada State Climatologist, July 1994.

7. Ibid.

8. Ibid.

9. Leland Sandquist, interview with author, 10 August 1994; Junior Cree, interview with Dennis Casebier, 31 October 1995.

10. James, Climatological Summary.

20: ACADEMY AWARD CHAMPION

1. *Searchlight,* 10 July 1903.

2. Ibid., 25 August 1905.

3. Head, *The Dress Doctor,* 23.

4. Ibid.

5. Anna Rothe, ed., *Current Biography: Who's New and Why* (New York: H. W. Wilson, 1945), 276.

6. Head and Calistro, *Edith Head's Hollywood,* 11.

7. Ibid., vii.

8. Head, *The Dress Doctor,* 27.

9. *Las Vegas Sun Magazine,* 17 October 1983, 5.

10. Edith Head, interview with author, 1972.

11. Head and Calistro, *Edith Head's Hollywood,* p. 14.

21: HOW BIG WAS IT?

1. Philip Johnston, "Searchlight in the Days Agone," *Westways Magazine,* June 1935, 8–9.

2. *Searchlight,* 25 August 1905.

3. Ibid., 31 August 1906.

4. Victor Morawetz, letter to E. P. Ripley, 9 March 1906.

5. *Bulletin,* 6 August 1909.

6. *Searchlight,* 31 August 1906.

7. Ibid., 22 July 1904.

8. Ronald M. James, Director of the Nevada State Office of Historical Preservation, letter to author, 6 April 1996.

9. *Bulletin,* 22 December 1911.

10. Phillip Earl, "Preserving the History of Searchlight: Old Nevada Mining Camp," *Pay Dirt Magazine,* August 1989, 8b.

11. Ibid.

12. Ronald M. James, "On the Edge of the Big Bonanza: Declining Fortunes and the Comstock Lode," 5 June 1995, 1–2.

22: FLYING HIGH AND FLYING LOW

1. Mary Ellen Sadovich, "Searchlight," *Las Vegas Review-Journal—The Nevadan,* 30 April 1967, 25.

2. Berlin, *King of Ragtime.*

3. Wade Cavanaugh, "M'Goo Makes You Wanna Dance," *Las Vegas Sun,* 5 August 1979, 10D.

4. Ibid.; see also Bill Vincent, "Rags Returns to Searchlight," *Las Vegas Review-Journal—The Nevadan,* 6 February 1977, 6.

5. *Las Vegas Sun,* 5 August 1989, 10D; see also Vincent, "Rags Returns."

6. *Bulletin,* 15 March 1907.

7. Irene Brennan, "Scott Joplin and His Searchlight Rag," *Las Vegas Review-Journal—The Nevadan,* 2 November 1975, 4.

8. *Las Vegas Review-Journal,* 27 April 1951.

9. *Bulletin,* 17 May 1912.

10. Sally Macready Wallace, interview with author, 15 July 1995.

11. Ibid.

12. Ibid.

13. Ed Dodrill, *Las Vegas Sun,* 10 January 1988; reprinted in *Nellis Air Force Base House Guide,* 18 November 1994, 12–13.

14. John A. Macready, "The Non-Stop Flight Across America," *National Geographic* 46 (July 1924): 1–92; see also Dodrill, *Las Vegas Sun.*

15. Macready, "Non-Stop Flight," 73.

16. G. C. Krohn, "T-2 Anniversary," *Armchair Aviator* 2, no. 6 (1972): 18–30.

17. Thomas, *Famous First Flight That Changed History,* 60.

18. Ibid.

19. *Journal of American History* (Summer 1955): 156.

20. Sally Macready Liston, "Of Those Who Fly," in Major L. R. Carastro, ed., *Air University Publication* (1972); reprinted by permission of *American Aviation Historical Society Journal* (Spring 1969).

21. Dodrill, *Las Vegas Sun.*

22. Haenszel, *Searchlight Remembered.*

23. Arda Haenszel, memorandum to Dennis Casebier, 30 November 1995.

24. Haenszel, *Searchlight Remembered,* 14–17.

25. Ibid.

26. Bill Kelsey, interview with Dennis Casebier, 2 January 1996.

27. Haenszel, *Searchlight Remembered,* 22–28.

28. Ibid., 32–33.

29. *Searchlight,* 4 September 1903.

30. Casebier, *Fort Pah-ute, California,* 11.

31. Ibid., 16.

32. Ibid., 21.

33. Dennis Casebier, "Uncle Sam's Bluecoats on the Mojave Road," *Las Vegas Review-Journal—The Nevadan,* 7 December 1986.

34. Ibid.

35. Casebier, *Fort Pah-Ute, California,* 30–31.

36. Paher, *Nevada Ghost Towns and Mining Camps,* 287.

37. Casebier, *Fort Pah-Ute, California,* 32–56.

23: THE CAMP WOULDN'T FAIL

1. *Searchlight Journal,* 12 December 1946.

2. Maude Frazier, as quoted in John Edwards Bray, Superintendent of Public Instruction, *State of Nevada Biennial Report of the Superintendent of Public Instruction* (Carson City: State Printing Office, 1915), 52.

3. Ella Day, interview with author, 25 August 1995.

4. Carl Myers, interview with Don Reid, 22 March 1995; Carl Myers, interview with Dennis Casebier, 10 August 1995.

5. Bill and Dorothy Douglas Kelsey, interview with Dennis Casebier, 2 January 1996.

6. Jeanne Sherman McCall, interview with Don Reid, 8 February 1995.

7. Bill Kelsey, interview with Dennis Casebier, 2 January 1996.

8. *Las Vegas Sun,* 10 January 1988.

9. Shirley Nellis, interview with author, 29 June 1995.

10. Ed Dodrill, *Las Vegas Sun,* 10 January 1988; reprinted in *Nellis Air Force Base House Guide,* 18 November 1994, 13.

11. Brian Greenspun, "Where I Stand," *Las Vegas Sun,* 26 December 1984.

12. Shirley Nellis, interview with author, 29 June 1995; J. Catherene Wilman, Ph.D., and Senior Airman James D. Reinhardt, *A Pictorial History of Nellis Air Force Base, 1941–1996* (Nellis AFB: Office of History, Hdqtrs. Air Weapons Center, Air Combat Command, 1996).

13. Vance Leavitt, interview with Dennis Casebier, 27 October 1995.

14. University of Nevada, Mackay School of Mines, Bulletin no. 62, 1965.

15. Bill and Dorothy Douglas Kelsey, interview with Dennis Casebier, 2 January 1996.

16. Tyler, *From the Ground Up,* 123.

17. Sergeant Sidney Meller, "The Desert Training Center and C-AMA," *Army Ground Forces Study* 15 (1946): 42–59.

18. Jane Reid, interview with Eleanor Johnson, 17 September 1980.

19. Carl Myers, interview with Dennis Casebier, 10 August 1995.

24: THE RENEGADE INDIAN

1. *Bulletin,* 11 November 1910.

2. Ibid., 23 December 1910.

3. *Las Vegas Age,* 10 December 1910

4. *Bulletin,* 9 December 1910.

5. Ibid., 10 March 1911.

6. Ibid., 31 March 1911.

7. Ibid., 17 March 1911.

8. Ibid., 29 April 1910.

9. Ibid., 17 March 1911.

10. *Las Vegas Age,* 14 January 1911, 26 March 1911.

11. *Bulletin,* 24 February 1911, 17 March 1911, 7 April 1911.

12. *Las Vegas Review-Journal,* 11 July 1948, 2b.

13. Ibid., 28 February 1940.

14. *Bulletin,* 22 November 1912; *Las Vegas Review-Journal,* 28 February 1940, 1.

15. *Las Vegas Age,* 25 January 1919, 1.

16. Jane Overy, letter to author, 5 May 1996.

17. Donna Jo Andrus, interview with author, 21 May 1996.

18. *Las Vegas Age,* 15 and 22 February 1919.

19. *Las Vegas Review-Journal,* 11 July 1948, 1b.

20. *Las Vegas Age,* 18 March 1919 and 3 May 1919, 1.

21. Ibid., 2 February 1935, 2.

22. *Las Vegas Review-Journal,* 21 February 1940, 1.

23. Ibid.

24. Roy Chesson, "The Man Who Ate Snakes Raw," *Las Vegas Review-Journal,* 27 February 1966, 26–27.

25. Roy Chesson, "The Search for Queho's Gold," *Las Vegas Review-Journal,* 6 March 1966, 26–27; *Las Vegas Review-Journal—The Nevadan,* 11 June 1978, 4J.

26. Elbert Edwards, oral statement, unknown date.

27. *Searchlight,* 23 June 1905.

25: SEARCHLIGHT SURVIVED THE WAR

1. Marge Marshall, interview with Dennis Casebier, 15 August 1995.

2. Lee Sandquist, interview with Dennis Casebier, 10 August 1995.

3. *Las Vegas Review-Journal,* 6 October 1947.

4. *Searchlight Journal,* 31 October 1946.

5. Ibid., 26 September 1946, 17 October 1946.

6. Ricca Pinnel, interview with author, 23 November 1994.

7. *Tonopah Times Bonanza,* 1 October 1948.

8. Hulse, *The Nevada Adventure,* 202.

9. Elaine Tanner, interview with author, 1 July 1995.

10. *Searchlight Journal,* February 1947.

11. Joyce Dickens Walker, letter to author, 23 April 1996.

26: THE PROMOTER

1. Hulse, *The Nevada Adventure,* 198–201.

2. Walter Wagner, "The Man Who Talked Riches out of Rocks," *True Magazine,* March 1964, 48.

3. *Mills vs. State Bar of Calif.,* 296 P. 280.

4. *Mills vs. State Bar of Calif.,* 6 Cal. 2d 565, 1936.

5. *State of California vs. Mills,* #65731.

6. Walter Wagner, "The Man Who Talked Riches out of Rocks," 48.

7. Larry Reid, interview with author, 14 October 1995.

8. Walter Wagner, "The Man Who Talked Riches out of Rocks," 48.

27: THE WORLD'S OLDEST PROFESSION

1. *Searchlight,* 14 July 1905.

2. *Bulletin,* 25 November 1910.

3. Terry Hudgens, interview with Don Reid, 18 July 1995.

4. Vincent, *Las Vegas Review-Journal—The Nevadan,* 9 January 1977, 3–4, 6 February 1977; see also *Las Vegas Sun,* 5 August 1979, 6.

5. Hudgens, interview with Don Reid, 17 July 1995.

6. Erlene Smith, interview with author, 28 May 1995; Carl Myers, interview with Dennis Casebier, 10 August 1995.

7. Junior Cree, interview with Dennis Casebier, 31 October 1995.

8. Lee Sandquist, interview with Dennis Casebier, 10 August 1995.

9. Marge Marshall Sandquist, interview with Dennis Casebier, 15 August 1995.

10. Cree, interview with Dennis Casebier, 31 October 1995.

11. Doc Brown, interview with Dennis Casebier, 11 November 1995, 23–26.

12. Ellen Pillard, "Legal Prostitution: Is It Just?" *Nevada Public Affairs Review* 2 (1983): 45.

13. Ibid.

14. Richard Symanski, "Prostitution in Nevada," *Annals of the Association of American Geographers* 64 (September 1974): 371.

15. Cree, interview with Dennis Casebier, 31 October 1995.

16. *Cunningham vs. Washoe County,* 66 Nev. 60, 1949.

17. Guy Rocha, Nevada State Archivist, statement, March 10, 1988.

18. Mike Belshaw and Ed Peplow Jr., *Historic Resources Study, Lake Mead Recreation Area, Nevada* (Denver: National Park Service, U.S. Department of the Interior, August 1980), 146.

19. Ibid.

20. Roberts, *Comprehensive History of the Church,* 4:171, 5:118.

21. National Park Service Orientation Booklet, compiled late 1970s.

22. Hulse, *The Nevada Adventure,* 228.

23. Casebier, *Early Days at Hart,* 21.

24. Bishop, *The Castle Mountain Story,* 10–11.

25. Ibid., 56–57.

28: CONCLUSION

1. Arrington, *Great Basin Kingdom,* 241.

SELECTED BIBLIOGRAPHY

Arrington, Leonard. *Great Basin Kingdom.* Cambridge: Harvard University Press, 1958.

Bancroft, Hubert. *History of Arizona.* History of the Pacific States of North America. Vol. 17. San Francisco: History Company of San Francisco, 1890.

——. *History of Nevada.* History of the Pacific States of North America. Vol. 20. San Francisco: History Company of San Francisco, 1890. Reprint, Reno: University of Nevada Press, 1981.

Berlin, Edward. *King of Ragtime.* New York: Oxford University Press, 1994.

Bishop, Bob. *The Castle Mountain Story.* Essex, Calif.: Tales of the Mojave Road Publishing Company, 1992.

Carlson, Helen. *Nevada Place Names.* Reno: University of Nevada Press, 1974.

Casebier, Dennis. *Fort Pah-ute, California.* Norco, Calif.: Tales of the Mojave Road Publishing Company, 1974.

——. *Mojave Road Guide.* Norco, Calif.: Tales of the Mojave Road Publishing Company, 1986.

——. *Guide to the East Mojave Heritage Trail: Needles to Ivanpah.* Norco, Calif.: Tales of the Mojave Road Publishing Company, 1987.

——. *Early Days at Hart.* Essex, Calif.: Tales of the Mojave Road Publishing Company, 1991.

——. *Goffs and Its Schoolhouse.* Essex, Calif.: Tales of the Mojave Road Publishing Company, 1995.

Elliott, Russell. *Nevada's Twentieth-Century Mining Boom.* Reno: University of Nevada Press, 1966.

Haenszel, Ardeth M. *Searchlight Remembered.* Norco, Calif.: Tales of the Mojave Road Publishing Company, 1988.

Head, Edith. *The Dress Doctor.* New York: Little, Brown, 1959.

Head, Edith, and Patty Calistro. *Edith Head's Hollywood.* New York: Dutton, 1983.

Hulse, James. *The Nevada Adventure.* Reno: University of Nevada Press, 1966.

Lingenfelter, Richard. *Steamboats of the Colorado.* Tucson: University of Arizona Press, 1978.

Lingenfelter, Richard, and Karen Rix Gash. *The Newspapers of Nevada: A History and Bibliography, 1854–1979.* Reno: University of Nevada Press, 1984.

Lord, Eliot. *Comstock Mining and Miners.* Berkeley: Howell-North, 1883.

May, Robin. *History of the American West.* New York: Simon and Schuster, Exeter Books, 1985.

Moody, Eric. *Masters of Capital.* Chronicles of America Series. Vol. 22. New Haven: Yale University Press, 1919.

Myrick, David. *Railroads of Nevada and Eastern California.* Vol. 2. Berkeley: Howell-North, 1963.

Ostrander, Gilman. *Nevada: The Great Rotten Borough.* New York: Knopf, 1966.

Paher, Stanley. *Nevada Ghost Towns and Mining Camps.* Berkeley: University of California Press, 1970.

Price, Captain George F. *The War of the Rebellion: Official Records of the Union and Confederate Armies.* Vol. 1, Part 1. Washington, D.C.: Government Printing Office, 1897.

Riggs, John. *Reign of Violence in Eldorado Canyon.* Third Biennial of the Nevada Historical Society. Carson City, Nev.: Carson City Printing Office, 1913.

Roberts, B. H. *Comprehensive History of the Church.* 6 vols. Provo, Utah: Brigham Young University Press, 1965.

Roske, Ralph. *Las Vegas: A Desert Paradise.* Tulsa, Okla.: Continental Heritage Press, 1986.

Rothe, Anna, ed. *Current Biography: Who's New and Why.* New York: H. W. Wilson, 1945.

Scrugham, James G. *History of Nevada: A Narrative of the Conquest of a Frontier Land.* Reno: American Historical Society, 1935.

Thomas, Lowell. *Famous First Flight That Changed History.* New York: Doubleday, 1968.

Townley, John. "Early Development of El Dorado Canyon and the Searchlight Mining Districts." *Nevada Historical Society* 11, no. 1 (Spring 1968): 1–14.

Tyler, R. N. *From the Ground Up.* New York: McGraw-Hill, 1948.

INDEX

Page references to maps and illustrations are printed in italic type.